T0280543

ADAM IN MYTH AND HISTORY

HARVARD SEMITIC MUSEUM PUBLICATIONS

Lawrence E. Stager, general editor
Michael D. Coogan, director of publications

HARVARD SEMITIC STUDIES

Jo Ann Hackett and John Huehnergard, editors

Syriac Manuscripts: A Catalogue — Moshe H. Goshen-Gottstein
Introduction to Classical Ethiopic — Thomas O. Lambdin
Dictionary of Old South Arabic — Joan C. Biella
The Poet and the Historian: Essays in Literary and Historical Biblical Criticism — Richard Elliott Friedman, Editor
The Songs of the Sabbath Sacrifice — Carol Newsom
Non-Canonical Psalms from Qumran: A Pseudepigraphic Collection — Eileen M. Schuller
The Hebrew of the Dead Sea Scrolls — Elisha Qimron
An Exodus Scroll from Qumran — Judith E. Sanderson
You Shall Have No Other Gods — Jeffrey H. Tigay
Ugaritic Vocabulary in Syllabic Transcription — John Huehnergard
The Scholarship of William Foxwell Albright — Gus Van Beek
The Akkadian of Ugarit — John Huehnergard
Features of the Eschatology of IV Ezra — Michael E. Stone
Studies in Neo-Aramaic — Wolfhart Heinrichs, Editor
Lingering over Words: Studies in Ancient Near Eastern Literature in Honor of William L. Moran — Tzvi Abusch, John Huehnergard, Piotr Steinkeller, Editors
A Grammar of the Palestinian Targum Fragments from the Cairo Genizah — Steven E. Fassberg
The Origins and Development of the Waw-Consecutive: Northwest Semitic Evidence from Ugaritic to Qumran — Mark S. Smith
Amurru Akkadian: A Linguistic Study, Volume I — Shlomo Izre'el
Amurru Akkadian: A Linguistic Study, Volume II — Shlomo Izre'el
The Installation of Baal's High Priestess at Emar — Daniel E. Fleming
The Development of the Arabic Scripts — Beatrice Gruendler
The Archaeology of Israelite Samaria: Early Iron Age through the Ninth Century BCE — Ron Tappy
A Grammar of Akkadian — John Huehnergard
Key to A Grammar of Akkadian — John Huehnergard
Akkadian Loanwords in Biblical Hebrew — Paul V. Mankowski

ADAM IN MYTH AND HISTORY

Ancient Israelite Perspectives on the Primal Human

by

Dexter E. Callender Jr.

EISENBRAUNS
Winona Lake, Indiana
2000

ADAM IN MYTH AND HISTORY
Ancient Israelite Perspectives on the Primal Human

by
Dexter E. Callender, Jr.

Printed in the United States of America

Library of Congress Cataloging-in-Publication Data

Callender, Dexter E., 1962–
Adam in myth and history : ancient Israelite perspectives on the primal human / by Dexter E. Callender Jr.
p. cm. — (Harvard Semitic studies ; no. 48)
Includes bibliographical references and index.
ISBN 1-57506-902-4 (cloth : alk. paper)
1. Human beings—Origin. 2. Adam (Biblical figure) 3. Man (Theology)—Biblical teaching. 4. Bible. O.T.—Criticism, interpretation, etc. I. Title. II. Series.
GN281.C343 2000
599.93′8—dc21

00-032135

The paper used in this publication meets the minimum requirements of the American National Standard for Information Sciences—Permanence of Paper for Printed Library Materials, ANSI Z39.48-1984.♾™

This work is dedicated to
Karla, Trey, and Charles,
to my mother, Joyce Grant Callender,
and to my father, Dexter E. Callender.

Contents

Acknowledgments

This work is a revised version of a Ph.D. dissertation completed in the Department of Near Eastern Languages and Civilizations at Harvard University in the Spring of 1995. I would like to thank Professors Jo Ann Hackett and Peter Machinist who directed the project and committee members Professors Paul D. Hanson and Lawrence E. Stager. Their guidance and scholarly insight enabled this work to come to fruition. Many thanks are due to Drs. John T. Fitzgerald, Marvin A. Sweeney, Daniel L. Pals, Stephen Sapp, David Kling, J. Joyce Schuld, William T. Dickens, Henry Green, Robert F. Moore, Harvey Siegel, Hugh R. Page, and Edwina Wright for their advice, assistance, and encouragement.

I would like to thank Professors Jo Ann Hackett and John Huehnergard for accepting this work for publication in the Harvard Semitic Studies Series and for their keen editorial guidance in the preparation of this manuscript. Any errors are solely the responsibility of the author.

I wish to express my appreciation to the Research Council of the University of Miami for providing research funding assistance through the Max Orovitz Summer Stipend in the Humanities and the Research Support Grant.

Finally, I wish to thank my wife, Karla, whose love and selflessness facilitated the completion of this book; and my young sons, Dexter III and Charles, who have patiently awaited the completion of this work.

Dexter E. Callender, Jr.
Coral Gables, Florida, summer 1999

Preface

To date there has been no full-length treatment of the biblical Adam, or "primal human," traditions in their ancient Israelite setting. Older studies prevalent at the turn of the century were vast in scope historically and cross-culturally. More importantly, such studies were routinely carried out in the interest of shedding light on concepts such as the "son of man" and the "messiah" — ideas that were of much greater importance in the later periods of early Judaism and Christianity, and ideas that captured the public and scholarly imagination to a much greater degree.

This monograph provides an historical-critical analysis of the relevant biblical traditions in their own right — an analysis that is sensitive both to the present literary context of the traditions and to their roots in the world of the ancient Near East. The difficulty in dating the various texts involved precludes establishing a purely diachronic account tracing the developmental history of such ideas. On the other hand, it invites judicious and disciplined synchronic forays tempered by the awareness of historical background. Of the manifold angles one could explore regarding these traditions, this study focuses on the way various tradents used the ideas, the underlying social significance shared by these traditions, and how that significance was realized in the cultural milieu of ancient Israel.

Abbreviations

AB Anchor Bible.

ABD *Anchor Bible Dictionary*. Edited by D. N. Freedman. 6 vols. New York: Doubleday, 1992.

ABL *Assyrian and Babylonian Letters Belonging to the Kouyunjik Collection(s) of the British Museum*. Edited by R. F. Harper. Chicago: University of Chicago Press, 1892–1914.

AHw *Akkadisches Handwörterbuch*. Edited by W. von Soden. 3 vols. Wiesbaden: Harrassowitz, 1985.

AIO *Archaeologia Iranica et Orientalis.*

AJSL *American Journal of Semitic Languages and Literatures.*

ANEP *The Ancient Near East in Pictures Relating to the Old Testament*. Edited by J. B. Pritchard. 2d ed. with supplement. Princeton: Princeton University Press, 1969.

ANET *Ancient Near Eastern Texts Relating to the Old Testament*. Edited by J. B. Pritchard. 3d ed. Princeton: Princeton University Press, 1969.

AOAT Alter Orient und Altes Testament.

ArOr *Archiv orientální.*

AS *Anatolian Studies.*

ASTI *Annual of the Swedish Theological Institute.*

Atr. *Atra-Ḫasīs: The Babylonian Story of the Flood*. W. G. Lambert and A. R. Millard. Oxford: Clarendon, 1969. See MS symbols in Lambert and Millard, pp. 40–41, for other citations with *Atr.* (e.g., *Atr.* G).

AUSS *Andrews University Seminary Studies.*

AV Authorized Version of the Bible.

BA *Biblical Archaeologist.*

BASOR *Bulletin of the American Schools of Oriental Research.*

Bauer-Leander Bauer, Hans, and Pontus Leander. *Historische Grammatik der hebräischen Sprache des Alten Testaments*. Halle: Niemeyer, 1922; repr. Hildesheim: Olms, 1962.

BDB Brown, F., S. R. Driver, and C. A. Briggs. *A Hebrew and English Lexicon of the Old Testament*. Oxford: Clarendon, 1907.

BHS *Biblia Hebraica Stuttgartensia*. Edited by K. Elliger and W. Rudolph. Stuttgart: Deutsche Bibelgesellschaft, 1977.

BN *Biblische Notizen.*

BSOAS *Bulletin of the School of Oriental (and African) Studies.*

BTB *Biblical Theology Bulletin.*

BZ	*Biblische Zeitschrift.*
BZAW	*Beihefte zur ZAW.*
CAD	Oppenheim, A. L., et al. *The Assyrian Dictionary of the Oriental Institute of the University of Chicago.* Glückstadt: Augustin, 1956–.
CBQ	*Catholic Biblical Quarterly.*
CQR	*Church Quarterly Review.*
CT	*Cuneiform Texts from Babylonian Tablets in the British Museum.*
CTM	*Concordia Theological Monthly.*
CTU	*The Cuneiform Alphabetic Texts from Ugarit, Ras Ibn Hani, and Other Places.* Edited by M. Dietrich, O. Loretz, and J. Sanmartín. Münster: Ugarit-Verlag, 1995.
DCH	*The Dictionary of Classical Hebrew.* Edited by D. J. A. Clines. Sheffield: Sheffield Academic Press, 1993–.
Deluge	The Sumerian Flood Story. See M. Civil, ed., in Lambert and Millard, *Atra-Ḫasīs (Atr.)*, 138–45, 167–72; Kramer, *ANET*, 42–44.
Dhorme	Dhorme, Edouard. *A Commentary on the Book of Job.* Translated by Harold Knight. Camden, N.J.: Thomas Nelson and Sons, 1967.
Driver	Driver, Samuel R., and G. B. Gray. *A Critical and Exegetical Commentary on the Book of Job.* ICC. Edinburgh: T. & T. Clark, 1921.
EnEl	*Enuma Elish;* Labat, Rene. *Le poème babylonien de la création.* Paris: Adrien-Maissonneuve, 1935.
ER	*The Encyclopedia of Religion.* Edited by M. Eliade. 16 vols. New York: Macmillan, 1987.
ExpTim	*Expository Times.*
Gilg.	*The Epic of Gilgamish.* R. Campbell Thompson. Oxford: Clarendon, 1930 (unless otherwise noted).
Gk	Greek.
GKC	*Gesenius' Hebrew Grammar.* Edited by E. Kautzsch. Translated by A. E. Cowley. 2d ed. Oxford: Clarendon, 1910.
Gordis	Gordis, Robert. *The Book of Job: Commentary, New Translation and Special Studies.* New York: Jewish Theological Seminary of America, 1978.
HALOT	Koehler, Ludwig, Walter Baumgartner, and J. J. Stamm. *The Hebrew and Aramaic Lexicon of the Old Testament.* Translated and edited under the supervision of M. E. J. Richardson. 4 vols. Leiden: E. J. Brill, 1994–1999. Translation of KB.
HAR	*Hebrew Annual Review.*
HAT	Handbuch zum alten Testament.
HKAT	Handkommentar zum alten Testament.
HNT	Handbuch zum neuen Testament.
HTR	*Harvard Theological Review.*
HUCA	*Hebrew Union College Annual.*

IB — *Interpreter's Bible*. Edited by G. A. Buttrick et al. New York: Abingdon-Cokesbury, 1951–1957.
ICC — International Critical Commentary.
IDB — *The Interpreter's Dictionary of the Bible*. Edited by G. A. Buttrick. 4 vols. Nashville: Abingdon, 1962.
IDBSup — *Interpreter's Dictionary of the Bible: Supplementary Volume*. Edited by K. Crim. Nashville: Abingdon, 1976.
Irén — *Irénikon*.
JAOS — *Journal of the American Oriental Society*.
JBL — *Journal of Biblical Literature*.
JCS — *Journal of Cuneiform Studies*.
JJS — *Journal of Jewish Studies*.
JNES — *Journal of Near Eastern Studies*.
JNSL — *Journal of Northwest Semitic Languages*.
JPOS — *Journal of the Palestine Oriental Society*.
JQR — *Jewish Quarterly Review*.
JRAS — *Journal of the Royal Asiatic Society*.
JSOT — *Journal for the Study of the Old Testament*.
JSS — *Journal of Semitic Studies*.
JTS — *Journal of Theological Studies*.
KAR — *Keilschrifttexte aus Assur religiösen Inhalts*. Edited by E. Ebeling. Leipzig: J. C. Hinrichs, 1915–1923.
KAT — Kommentar zum Alten Testament.
KB — Koehler, Ludwig, Walter Baumgartner, and J. J. Stamm. *Hebräisches und aramäisches Lexikon zum Alten Testament*. 3d ed. Leiden: E. J. Brill, 1967–1995.
Lambdin — Lambdin, Thomas O. *Introduction to Biblical Hebrew*. New York: Scribner's, 1971.
LKA — *Literarische Keilschrifttexte aus Assur*. Edited by E. Ebeling. Berlin: Akademie-Verlag, 1953.
LSJ — Liddell, H. G., R. Scott, and H. S. Jones. *A Greek-English Lexicon*. 9th ed. with revised supplement. Oxford, 1996.
LXX — Septuagint: Greek translation of the Hebrew Bible.
MT — Masoretic Text (received text of the Hebrew Bible).
NRSV — New Revised Standard Version of the Bible.
NTT — *Norsk Teologisk Tidsskrift*.
Or — *Orientalia*.
OTL — Old Testament Library.
OTP — *Old Testament Pseudepigrapha*. Edited by J. H. Charlesworth. 2 vols. New York: Doubleday, 1983.
Perf — Perfect form of the Hebrew verb.
Pope — Pope, Marvin. *Job*. AB. Garden City, N.Y.: Doubleday, 1973.

Q Qumran manuscript.
RA *Revue d'assyriologie et d'archéologie orientale.*
RB *Revue biblique.*
RHR *Revue de l'histoire des religions.*
RO *Rocznik Orientalistyczny.*
RV Revised Version of the Bible.
Sam Samaritan Hebrew version of the Hebrew Bible.
SE *Svensk exegetisk ārsbok.*
SEL *Studi epigrafici e linguistici.*
SJOT *Scandinavian Journal of the Old Testament.*
SJT *Scottish Journal of Theology.*
ST *Studia theologica.*
STK *Svensk teologisk kvartalskrift.*
Syr Syriac translation of the Hebrew Bible.
TDOT *Theological Dictionary of the Old Testament.* Edited by G. J. Botterweck and H. Ringgren. Translated by John T. Willis. Grand Rapids : Eerdmans, 1974–.
Tg Targum: Aramaic translation of the Hebrew Bible.
TLZ *Theologische Literaturzeitung.*
TWAT *Theologisches Wörterbuch zum Alten Testament.* Edited by G. J. Botterweck and H. Ringgren. Stuttgart: Kohlhammer, 1970–.
TZ *Theologische Zeitschrift.*
UF *Ugarit-Forschungen.*
UVB Vorläufiger Bericht über die von der Notgemeinschaft der Deutschen Wissenschaft in Uruk-Warka unternommenen Ausgrabungen.
VT *Vetus Testamentum.*
VTSup Vetus Testamentum Supplements.
Vulg Vulgate.
WMANT Wissenschaftliche Monographien zum Alten und Neuen Testament.
ZAW *Zeitschrift für die alttestamentliche Wissenschaft.*
Zimmerli Zimmerli, W. *Ezekiel.* 2 vols. Translated by James D. Martin. Hermeneia. Philadelphia: Fortress, 1983.

Introduction

I

THE BIBLICAL FIGURE OF ADAM has captured the imagination of Western culture for millennia. His story has been told and retold repeatedly in the works of Milton, Shelley, Steinbeck, and countless others. That he was created, as Genesis 2–3 reports, from dust of the earth, placed in the garden of Eden, witnessed the creation of woman, partook of a mysterious forbidden fruit with that woman, and was subsequently expelled from the garden as a result is a tale so familiar that we hardly need to summarize it here.

But this is not the only tradition concerning humanity in its dawn that the Bible preserves. The creation narrative in Genesis 1 is known for its famous directive issued by the deity — the curious and oft debated 'Let us create a human in our image, according to our likeness' (1:26) — and the accompanying summary statement:

> So God created humanity in his image:
> In the image of God he created him,
> Male and female he created them. (1:27)[1]

Still, there are other biblical traditions that are less heralded. In the book of Job, the anguished and bewildered hero ponders the possible causes of his wretched fate, while well-meaning friends chide him with their own brand of "wisdom." In 15:7–9, one of those friends, Eliphaz, appears to lose patience with the sufferer and blurts:

> Were you born the first human,
> And brought forth before the hills?
> Did you listen in the council of God,
> And take wisdom for yourself?
> What do you know that we know not?
> What do you understand that escapes us?

[1]All translations are the author's, unless otherwise noted.

These words of Eliphaz betray a different understanding of the first human on the part of the narrator from what we saw previously in Genesis. The figure is born 'before the hills' and stands 'in the council of God'. We may also note here that the context in which the reference appears is quite different.

Yet another tradition — one that echoes the themes of the Genesis 2–3 account in a rather ambiguous fashion — is set within a prophetic oracle against the king of Tyre, attributed to the sixth century exilic period prophet Ezekiel. In 28:11–19 the prophet raises a lament over the king of Tyre, whom he likens to a mythical being. Ezekiel describes the being as 'full of wisdom and perfect in beauty'. He asserts that the figure was 'in Eden, the garden of God' and 'on the holy mountain of God'. Then, according to the prophet, something went awry:

> You were pure in your ways, from the day you were created,
> Until iniquity was found in you. (v. 15)

He goes on to say that the figure was expelled from the mountain of God and ultimately 'turned to ashes'. Despite grammatical, textual, and interpretive difficulties, this too certainly suggests a figure akin to the first humans featured in the Genesis 2–3 account. The ideas are shaped differently and, again, they are used in a different context.

Common Elements

The figure presented in each of the texts is essentially human, belongs to the primeval era, and has unusual contact with the divine realm. Further, other common elements present a useful focal point for a study of these texts.

I have chosen to identify this figure quite simply as "the primal human," a designation that highlights the primordial setting of the figure and its essential humanity. I have found the use of the term "man" in this context to be misleading for many readers, given the general awareness of gender issues in language. The apparent absence of any emphasis on the masculinity of the figure and the regular use of the generic term *ʾādām* further commends the appellation "human." As for the often applied German term *Urmensch,* it smacks of jargon and has, as we shall see, acquired a technical significance in some circles, making it a less than appropriate option.

The term "primal" reflects the observation that the figure is connected with *creation*, the beginnings of the cosmos.[2] The connection to creation may be understood in many ways, but among the more significant is that it represents what scholars of religion have come to express as mythical or sacred time. M. Eliade popularly characterized this important foundational epoch by the phrases "once upon a time" and "*in illo tempore, ab origine.*"[3] In this regard, the designation "primal" seems appropriate, in the sense that the figure is understood to belong to the period of first or foundational things.

Further comparison of the texts outlined above reveals other elements in addition to *creation* that the texts share, at least the texts in which the figure is presented as an actor. Thus, with the exception of Genesis 1, three such elements may be identified as *location, wisdom,* and *conflict.*[4] To various degrees, each of these elements, or *topoi,* plays a part in establishing the conception. With regard to *location,* Genesis 2–3, Job 15, and Ezekiel 28 express an interest in the place where the primal human acts. In each case location is special. In Genesis, the garden of Eden is the locus into which the primal human is 'placed'. Likewise, in Ezekiel 28, the primal human is set in Eden, 'the garden of God'. In Job 15, the action of the primal human occurs within the context of 'the council of God' (*bəsôd ʾĕlōhîm*). Although each text expresses it in different words, each places the primal human within what is a locus of divine activity.

As for the topos *wisdom,* in the Genesis 2–3 narrative, the concept of wisdom is embodied in the tree of the knowledge of good and

---

[2]The context of creation in Genesis is obviously central. In Ezek 28:11–19, although creation does not play a major role, the text makes reference to the creation of the protagonist. The reference to the first human in Job 15:7 speaks of his "birth" before the hills.

[3]M. Eliade, *Cosmos and History: The Myth of the Eternal Return* (New York: Harper and Row, 1959), 21.

[4]Gen 1:26–28 presents the first human in a different manner from the other texts insofar as no reference is made to any type of action carried out by the figure (unlike Job 15:7–8, and Ezek 28:11–19).

evil and the well-known incident that surrounds it. In Job 15 the primal human is introduced into the discussion explicitly on account of his association with wisdom. Likewise wisdom plays a major role in the course of action followed by the primal human in Ezek 28:11–19.[5] Biblical scholars have by no means overlooked this element, although a consensus regarding its significance has yet to emerge.[6]

Finally, each of these texts presents tension or potential conflict that is related to the location and wisdom of the primal human. In Genesis, eating the fruit from the tree of the knowledge of good and evil results in the primal human's expulsion from the garden. In the second oracle of Ezekiel 28, the corruption of wisdom leads to the same. In Job 15 such conflict is alluded to in the general statement, 'God places no trust in his holy ones' (v. 15). It may also be expressed in the ambiguous language, which I have translated, 'did you take wisdom for yourself?' (v. 8). Many have understood this conflict as a result of what may be, for now, termed "hubris," a notion which is in the most basic sense a reflection upon the relationship between human and divine; it is an overstepping of humanity's bounds. We will explore the place of creation, location, wisdom, and conflict in the chapters to follow.

---

[5]Note also the personified expression of wisdom in Proverbs 8.

[6]Both Feuillet and Muilenburg noted the importance of wisdom in primal human traditions, considering the figure a hypostasis of wisdom. A. Feuillet, "Le fils de l'homme de Daniel et la tradition biblique," *RB* 60 (1953): 170–203, 321–46; J. Muilenburg, "The Son of Man in Daniel and the Ethiopic Apocalypse of Enoch," *JBL* 79 (1960): 197–209. Note the contrary opinion of J. Coppens, "Le messianisme sapiential et les origines littéraires du fils de l'homme daniélique," *Wisdom in Israel and the Ancient Near East* (ed. M. Noth and D. W. Thomas; VTSup 3; Leiden: E. J. Brill, 1969), 33–41. Habel sees the image of the primal human operating alongside that of a personified form of Wisdom (N. C. Habel, *The Book of Job: A Commentary* [OTL; London: SCM Press, 1985]). See also W. Schencke, *Die Chokma (Sophia) in der judischen Hypostasenspekulation : Ein Beitrag zur Geschichte der religiosen Ideen im Zeitalter des Hellenismus* (Kristiania: Utgit for H. A. Benneches Fond, in Kommission bei J. Dybwad, 1913),15–25. Studies comparing Adam and the Mesopotamian figure Adapa generally recognize wisdom as an important element. See J. Pedersen, "Wisdom and Immortality," *Wisdom in Israel and the Ancient Near East*, 238–46.

Related Passages

There are several other passages in the Hebrew Bible that employ language which is strongly reminiscent of imagery presented in the texts already mentioned, yet it is immediately apparent that they are not *simple* references to the same figure. The images found in these texts are, as we shall see, appropriately considered vestigial allusions to the primal human. Such texts draw upon and are embellished by stock imagery of the primal human. Chief among these are Ezek 28:1–10 and Prov 8:22–31. Echoes of the tradition are likewise apparent in Psalms 8 and 82.

Where do we go from here? Given the similarities and differences in content and presentation, we are led to wonder what an ancient Israelite audience might have made of such connected, yet varied traditions. How was this imagery used by the various authors who appeal to it, and what might an ancient reader have understood to be implicit in the imagery? Such questions embody the essential problem this volume proposes to address.

II

The similarities and differences among these texts have certainly not escaped the attention of biblical scholars. But surprisingly, in spite of the interest that these texts have generated, to date no one has produced a full-length study that investigates the background and interrelations of the texts. The work that has been done has provided a number of insights, but has largely been of a cursory nature. For convenience, we might consider such work as falling roughly into three periods or movements: that of the History of Religions School around the turn of the century, the Myth and Ritual School in the years surrounding World War II, and finally, more recent studies of the past four decades.

The History of Religions School

The work of the so-called *religionsgeschichtliche Schule* in Germany represented the first efforts of modern scholarship to address these texts, and the interests that defined the group are evident in the manner by which they approached them. The school originated with a group of Protestant German scholars who in the late nineteenth century sought to apply a comparative history of religions method to the Bible. Most of the effort was focused ultimately on the New Testament, in particular on elucidating the nature and background of early Christianity. Hence, the primal human texts were explored only as elements which contributed to a later picture. The greatest legacy of the school, especially with respect to the issue at hand, was to introduce the comparative method to the inquiry.

The earliest statements appeared in the work of Hermann Gunkel, one of the founding members of the *religionsgeschichtliche Schule,* and the most widely celebrated. His well-known study of biblical beginning and end-times traditions, *Schöpfung und Chaos in Urzeit und Endzeit,* was published in 1895 and focused on Genesis 1 and Revelation 12. Its subtitle appropriately defined it as a "history of religions inquiry." Gunkel brought Ezekiel 28 and Job 15 into his investigation of Genesis, speaking of them as "two allusions to an older recension of the Paradise History."[7] His comments here, however, were presented in the most general of terms, limited to a cursory description of each text. In keeping with his comparative concerns, Gunkel presented the Mesopotamian figure Adapa as a reflection of the older recension. His ultimate goal in this brief survey was to demonstrate the Babylonian origin of the traditions. In doing so, Gunkel wrote of the "enormous power of Babylonian culture in these lands so widely removed from Babylon" and went on to assert that "we can accept that Syro-Palestinian culture was completely saturated by Babylonian."[8] Gunkel drew his conclusions with little discussion, given the fact that this was background to his greater work. Later, in his commentary on Genesis, Gunkel elaborated on his comments, but not by much. In it, he described the passages at greater length and gave slightly more attention to specific elements of comparison. He

---

[7]Gunkel, *Schöpfung und Chaos in Urzeit und Endzeit: eine religionsgeschichtliche Untersuchung über Gen 1 und Ap Joh 12* (Göttingen: Vandenhoeck und Ruprecht, 1895), 148.

[8]Ibid., 151.

even recognized the literary setting of the allusions, but again, his comments were limited.[9]

A few short years after *Schöpfung and Chaos*, Wilhelm Bousset, another of the school's founders, similarly commented on these traditions. Bousset, a New Testament scholar, was primarily interested in the relationship of the Hellenistic religions to early Judaism and Christianity. In his most notable work, *Die Religion des Judentums im späthellenistischen Zeitalter* (1926), Bousset considered more closely another aspect of the variants — that among the variant biblical passages, the protagonist was either more or less "exalted." Bousset's interest in Hellenistic contributions led him to conclude that the protagonist seen in passages such as Ezek 28:12–19, and Job 15:8 was a god-like primal human (*göttergleicher Urmensch*) comparable to the exalted forms of Adam found in early Jewish literature (citing as the first among many examples *2 En.* 30:6–14).[10] Bousset considered the exalted nature of the figure, however, to hail not from within Judaism, but from a foreign idea that became fused with the original Jewish idea. This work was primarily in the service of Bousset's interest in the concepts "messiah" and "son of man," both of which he believed were fed by outside forces, Gnostic, Mandaic, Manichaean, and Persian among others.

Hugo Gressmann, who became associated with the group shortly after the turn of the century, addressed the texts in his *Ursprung der israelitisch-jüdischen Eschatologie*, published in 1905. Gressmann's efforts to cast light on the eschatological messianic king led him to biblical references made to paradise or paradise conditions, such as those found in Isaiah 7, 9, 11, and Micah 5. True to his comparative interests he found a key to the origins of the idea in the Indian figure Yama (Iranian Yima). Although Gressmann essentially followed Gunkel's reconstruction of an original myth, he offered more detailed points of comparison and more discussion of the texts, although the discussion was limited to matters of basic content. Like Bousset and Gunkel, both

---

[9] *Genesis*. Vol. 1. (HKAT; Göttingen: Vandenhoeck und Ruprecht, 1910).

[10] *Die Religion des Judentums im späthellenistischen Zeitalter* (HNT; Tübingen: J. C. B. Mohr, 1926). Bousset also mentioned Prov 30:4 as reflecting the same traditional material (352).

of whom he cited at various points in the study, Gressmann considered the Adam narrative of Genesis 2–3 to reflect a distinct tradition, essentially unrelated to the other materials that are similar in appearance.

Shortly after Gressmann published *Ursprung*, the ideas of the *religionsgeschictliche Schule* were echoed in England by T. K. Cheyne, with the completion in 1905 of his *Traditions and Beliefs of Ancient Israel*.[11] Cheyne intended the study to reflect the advances brought about by the growth of Assryiology and the discovery of new Mesopotamian texts which promised to shed light on the biblical materials. It essentially presented his ideas on the traditions of the books of Genesis and Exodus, some of which he had earlier hoped to present in a two-volume commentary proposed for the International Critical Commentary Series.[12] As such, the book represented the first thoroughgoing investigation of several Genesis traditions from a comparative perspective.[13] Cheyne's treatment of the primal human traditions did not differ greatly from the opinions already expressed by Gunkel, Bousset, and Gressmann. He too appealed to the Avestan tradition of Yima "the first king and founder of civilization," as relevant for understanding Ezek 28:11–19 and Job 15:7–9. He similarly understood these references along with Gen 1:26 to reflect an older form of the tradition that was later modified in a form now preserved in Genesis 2–3.

The comparative work begun by the German *religionsgeschichtliche Schule* was carried out in America in the work of C. H. Kraeling. In *Anthropos and Son of Man*, Kraeling continued earlier work which focused on the related mythical figure known in Western Gnostic systems as the "Anthropos." In exploring the biblical materials, however, he came to different conclusions regarding the relationship of Genesis 2–3 to the variant traditions in Ezekiel 28 and Job 15. Kraeling considered Genesis 2–3 to reflect the older tradition — a preexilic view of humanity that was essentially "pessimistic" and that con-

---

[11]The work was ultimately published in 1907 (London: Adam and Charles Black).

[12]*Traditions and Beliefs*, v–vi.

[13]Cheyne appealed to various cultures to illuminate the biblical text, including Finnish and Native American traditions.

tinued in the condemnatory views of Adam found, among other places, in *4 Ezra* (3:7; 4:30; 7:11) and *2 Baruch* (23:4; 48:42). These passages blame the woeful state of humanity on the first man. An exalted understanding of Adam and a true conception of a Primal Man came with the exile and is evidenced in Ezek 28:12–19, the P account of the creation of humanity (Gen 1:26), Job 15:7–9, and Psalm 8. Kraeling argued that these texts lie behind the laudatory views of Adam in the later literature, which shift blame for the Fall (the curse subsequent to the disobedience in the garden) onto others (Wis 10:3, Cain; *Jubilees* 5, demons; *1 Enoch* 6, angels), and present him as highly exalted (Sir 49:14–16; *2 Enoch* 8; 2 Cor 12:3–4).[14]

This brief survey of Gunkel and like scholars demonstrates how the *religionsgeschichtliche Schule* approached the different ideas presented in the texts in question. The studies were generally interested in understanding later phenomena, such as son of man and messiah, which limited the amount of discussion that could be devoted to the biblical texts themselves. Still, the aggressive emphasis on comparative analysis, though scaled back in later years, demonstrated the extent to which Israel was in fact not *sui generis*, but a part of its cultural environment. The concern for comparative analysis is a legacy of the school which survives to this day, and remains an important part of producing an understanding of the texts which concern the present volume.

The Myth and Ritual School

A different emphasis in the study of primal human traditions may be discerned with the rise of the so-called Myth and Ritual School, a term which refers to two movements in the second quarter of the twentieth century in Great Britain and Scandinavia, respectively.[15] The common threads that joined these scholars was an interest in the cen-

---

[14]C. H. Kraeling, ***Anthropos and Son of Man: A Study in the Religious Syncretism of the Hellenistic Orient*** (New York: Columbia University Press, 1927).

[15]This approach began with the work of classicists and was adopted by biblical scholars. In Great Britain, the biblical movement was spearheaded by S. H. Hooke, while in Scandinavia, biblical practitioners included (among others) G. Widengren, I. Engnell, S. Mowinckel, and A. Bentzen.

tral importance of ritual acts among ancient peoples and the conviction that myths were the texts that accompanied such acts. A point of continuity with earlier commentators of the *religionsgeschichtliche Schule* was to approach the texts in question within the scope of inquiry into the messianic concept. This interest was manifested largely in the idea of sacral kingship, the theory that in many ancient Near Eastern cultures the king was considered to be divine and as such was a participant in the cultic drama.[16]

Typifying this emphasis with respect to the primal human was Aage Bentzen's *Messias — Moses Redivivus — Menschensohn,* translated and published in English in 1955 under the title *King and Messiah.*[17] This short monograph was an investigation of sacral kingship in ancient Israel and issues in the interpretation of the psalms, the messianic hope, the suffering servant of Yahweh, and the son of man. In his discussion of Psalm 2 and "the Messiah in the Psalms," Bentzen found the Israelite primal human to be an idea essentially associated with kingship. The association manifested itself in the annual enthronement festival — a cultic ceremony suggested in the royal psalms (e.g., 2, 89, 110) celebrating the rule of the king. Bentzen argued that in this festival, the enthronement of the Israelite king repeats the primordial enthronement of the mythical "first king," who is "the patriarch of the Royal House, identical with the patriarch of mankind."[18] He found support for this contention in passages such as Mic 5:1, according to which the royal figure comes 'from the days of old' (*miqqedem*), and Isa 9:5, where the king is called *ʾăbî-ʿad* 'Father from Eternity'. Bentzen classified these prophetic passages as "typical royal psalms" and concluded that the king *is* "Primeval Man."[19] He went further in asserting that the royal imagery of Gen 1:26–28, Psalm 8, and Genesis 2 is based upon the notion of the king as "Primeval Man."[20]

---

[16]Ivan Engnell developed and articulated this position in *Studies in Divine Kingship in the Ancient Near East* (Oxford: Basil Blackwell, 1967).

[17]See the generally available edition, Oxford: Basil Blackwell, 1970.

[18]*King and Messiah,* 17.

[19]Ibid.

[20]Bentzen (18) followed G. Widengren in interpreting the superiority of the first man over the animals and his naming them as an expression of royal imagery in

Bentzen also brought Ezek 28:1–10, 11–19, and Job 15:7–8 into the discussion as "other descriptions of First Man in the Old Testament." He likewise asserted that these texts, especially Ezekiel, "point to a king."[21] Bentzen, however, never explained how the passage in Job 15 is related to kingship; in fact, throughout the entire work, he never again discussed Job 15. Bentzen's discussion of Ezekiel 28 is similarly cursory. He related the Ezekiel oracles to the son of God in Psalms 2 and 10, who lives 'on the "Mountain of the Gods"'.[22]

Adherents of the Myth and Ritual school also addressed concerns arising from the comparative work of earlier commentators. Undoubtedly the most influential proponent of the myth and ritual relationship was Sigmund Mowinckel, whose *He That Cometh* — a study of the messiah concept — touched on the texts in question.[23] Like those of Bentzen and others, Mowinckel's discussion of the texts is cursory; there is no extensive discussion of Job 15. His myth and ritual orientation is clear in his comments on Ezekiel 28, asserting that enthronement was repeated as an annual festival, a reenactment of creation which reestablished fertility of every kind. "It was therefore quite natural that the installation of the king and the New Year festival should be regarded as the preservation and re-creation of the primeval splendour. To the Israelite, all the glory of the earth was summed up in the thought of 'Yahweh's garden,' the 'garden of God,' told of in the ancient creation myths."[24] Mowinckel also attempted to clarify the comparative issue by asserting that a distinction should be made between the first man and the *Urmensch*, a term typically used to describe figures such as Indo-Iranian Gayomard and Ymir of Scandinavian tradition.[25] His understanding of the distinction was that the Urmensch was "in its nature a cosmological idea, i.e., it originated as explanation of the origin of the cosmos. The Urmensch is, in this

---

Genesis 2. This, however, was the extent of his evidence.

[21]Ibid., 18; cf. 42, where he makes the same assertion in more emphatic terms.

[22]Ibid., 42.

[23]Translated by G. W. Anderson (New York: Abingdon, 1954); first published in 1951 under the title *Han som kommer*.

[24]Ibid., 55.

[25]Ibid., 423–24.

regard, the cosmos itself, put forward in human form, macrocosm put forth as microcosm in human form."[26] In contrast, the first created man is simply the progenitor of the human race, and not a "speculative cosmological idea."[27]

Following Mowinckel's rejection of Bentzen's association of the first man with foreign Urmensch figures, Bentzen modified, but defended his position, stating that while there is no expression of the micro- and macrocosms typified in the Scandinavian Ymir and Iranian Gayomard, the first man of Genesis 1–2, Ezekiel 28, and Job 15 "has a meaning and a significant position among the ideas of the creation of the world." "The Adam of the Old Testament," he concluded, "is a variant of the idea of 'Primordial Man'" — the Urmensch.[28] Bentzen's comments demonstrated the extent to which the nature of the Israelite conception of the primal human and how it relates to similar ideas from other cultures remained unclear.

Despite the fact that proponents of the Myth and Ritual school succeeded in highlighting the place of kingship in certain biblical texts, they were not interested in how the imagery was put to use in the various contexts. Nor did they ask, in turn, how those uses might inform our understanding of the figure.

Recent Studies

The last four decades show continued interest in the primal human traditions. It is clear, however, that treatment of the texts has continued to be cursory or within the discussion of other phenomena.[29] During this period, two studies have appeared that may be distinguished from earlier studies in the extent to which they focus on the pertinent texts as a whole. These are H. May's "The King in the Garden of Eden" and D. Gowan's thematic study of humanism and hubris,

---

[26]"Urmensch und Königsideologie," *ST* 2 (1948): 71.

[27] Ibid., 72.

[28]*King and Messiah*, 43.

[29]See, for example, the treatment of J. Van Seters, "The Creation of Man and the Creation of the King," *ZAW* 101 (1989): 333–42, which does not consider the allusion in Job 15:7–8.

When Man Becomes God.[30]

Herbert May used Ezek 28:11–19 to reconstruct what he called the "basic underlying pattern" of a myth of a "royal first man" upon which were based the allusions in Psalm 8, Prov 8:22–31, Job 15:7,8, and 38:4–7, and Daniel 7.[31] May considered the references to the son of man in the Similitudes of Enoch (*1 Enoch* 37–71) and 4 Esdras also to have been "influenced by the motif of the pre-creation royal First Man in the wisdom literature."[32] May's brief study is significant in that he brings together the critical mass of primal human traditions in the Hebrew Bible. He too finds the royal aspect to be significant. He also highlights some of the key scholarship which collectively assembles these passages. It does not, however, appear to have been his purpose to discuss the various contexts in which this imagery appears.

D. Gowan has brought together several of the texts considered here and has wrestled with the meaning of the primal human traditions in his thematic study on hubris, *When Man Becomes God: Humanism and Hybris in the Old Testament*. The work originated as an inquiry into the popular genre of prophetic literature commonly referred to as "oracles against the nations." It ultimately became an exposition of how hubris, which he defines as full rebellion against God, manifests itself in the human political state — that aspect of humanity that can legitimately threaten to "replace" God. Gowan's work bears relevance for our study insofar as he underscores the connection of equality with God, defiance of God, and subsequent punishment to the traditions which form the basis of our study. His interest in the oracles against the nations paves the way for further exploration of the literary contexts of the passages and reminds us how attention to literary context can prove useful for shedding light on the ideas embodied in these passages.

---

[30]H. May, "The King in the Garden of Eden," *Israel's Prophetic Heritage* (ed. B. W. Anderson and W. Harrelson; New York: Harper, 1962), 166–76; D. Gowan, *When Man Becomes God: Humanism and Hybris in the Old Testament* (Pittsburgh Theological Monograph Series 6; Pittsburgh: Pickwick, 1975).

[31]For a similar discussion, see F. H. Borsch, *The Son of Man in Myth and History* (London: SCM Press, 1967), 89–131.

[32]"The King in the Garden of Eden," 174–75.

Three observations regarding Gowan's treatment suggest room for further work. First, Gowan is interested in pursuing a theme, the scope of which is much larger than the texts proposed here for study. What is more, the meaning resident in the texts of our smaller corpus extends beyond Gowan's theme and is by no means exhausted by it. In this regard, his study may be classified along with the works of the history of religions school for which the same might be said. Second, he gives little attention to the concept of wisdom, which his silence suggests is treated as an incidental attribute of the protagonist.[33] We shall see that the definition and nature of wisdom are of central importance in these texts. Finally, Gowan sets up the problem in essentially negative terms: humanity versus God or humanity replacing God. By taking this negative approach, he does not explore the positive aspects of the human/divine relationship — an aspect recognized by earlier scholarship as mentioned above.[34]

III

Genesis 1; 2–3; Job 15:7–8; and Ezek 28:11–19 have understandably been central to scholarly concern with Israel's primal figures, although, again, they have not been extensively examined together. To date, no study exists that is given specifically to the *systematic* investigation of these texts, their significance, and how they relate to one another. This, and the lack of attention on the part of earlier commentators to matters such as literary context, has created the need for a fresh examination. The lack of consensus regarding the nature of the protagonist in each text and indeed the lack of *any* clear statement of how these texts relate to one another highlight the need for a volume that addresses such issues. Such a situation is perhaps most evident in discussions found in general reference works — dictionaries and encyclopedias, where mention of certain of these texts as variant

[33]Ibid., 87. Gowan recognizes wisdom as a characteristic feature among ancient Near Eastern and biblical "prototypes of humanity"; however, he appears to treat it as a self-evident concept, by giving no discussion to it.

[34]Cf. discussion above of C. H. Kraeling, who found the presence of a positive aspect that formed the basis for the laudatory views of Adam of later literature.

expressions of a common tradition is conspicuously absent.[35] These discussions are strangely silent with regard to Job 15:7–8 and Ezek 28:11–19, suggesting that these texts are not to be regarded in the study of the ancient Israelite concept of Adam. Common sense, however (particularly given the reference to the "first human" in Job), and the opinion of scholars outlined above dictate otherwise, even though a fair assessment of such silence acknowledges the difficulties with which these allusions are fraught.

What Remains to Be Done

What I hope to accomplish, then, in this treatment is twofold. First, I will highlight the essential characteristics uniting these biblical traditions into a coherent set of ideas, which constituted a symbol that was recognized by the ancients and not just by modern critics. This study sets out to identify the conceptual nucleus of such traditions, which I have identified as the symbol of the "the primal human." Second, I will establish how this "primal human" symbol functioned in each of its literary contexts, to forge an understanding of its various manifestations and permutations. What did the use of the figure mean

---

[35]In his general introductory article "Adam" in the *Interpreter's Dictionary of the Bible*, Brevard Childs devotes no space to the subject. He discusses Adam in "the Genesis Narratives" as well as in "non-canonical Jewish literature," but does not mention the traditions in Job or Ezekiel (B. Childs, "Adam," *IDB* 1.42–44). Likewise, more recently Howard Wallace's general article on Adam in the *Anchor Bible Dictionary*, despite a rich presentation of textual references to *ʾādām* in Genesis 1–11 and "in intertestamental literature," and brief allusions to Jewish traditions concerning the cosmic proportions of Adam and his great wisdom (*Gen. Rab.* 8:1; 21:3; 24:2; *Pirqe R. El.* 11; *ʾAbot R. Nat.* B8, etc.), makes no reference to the traditions of Job or Ezekiel. (H. Wallace, "Adam," *ABD* 1.62–64.). Michael Fishbane's discussion of Adam in the *Encyclopedia of Religion* makes reference to Adam Qadmon of the Qabbalah. He does not, however, mention the possibility or likelihood that at least a variation of the figure appears elsewhere in the Hebrew Bible (M. Fishbane, "Adam," *ER* 1.27–28. The same holds true in general discussions of Messiah; see M. de Jonge, "Messiah," in *ABD* 4.777–88; E. Rivkin, "Messiah, Jewish," *IDBSup* 588–91. For discussions of Son of Man, see N. Perrin, "Son of Man," *IDBSup* 833–36; and S. E. Johnson, "Son of Man," *IDB* 4.413–20. In G. W. E. Nickelsburg's "Son of Man," (*ABD* 5.137–50) one finds the single observation that Psalm 8 is a combination of Gen 1:26–28 and Ezek 28:12–18.

to the writers, hearers, redactors, and tradents of ancient Israel? What was the significance of invoking the imagery? It is the aim of this study to provide an account of how these traditions were understood in ancient Israel and how they were used — a task that has yet to be undertaken.

As for the method and plan of this work, I will address these issues through a close reading of each text in question — in the light of cognate and comparative literature from around the ancient Near East. The focal point will remain the biblical traditions, but comparative materials will be prudently scrutinized as appropriate. Because these traditions developed over time, diachronic historical issues must be addressed. The problems inherent in dating traditions within the biblical text, however — particularly with respect to the passages of concern here — limit what one might assert with confidence. Because the various images are closely related and because they ultimately came to reside within a single text — the canon — synchronic analysis is equally important. Given the problem of dating, synchronic analysis will be emphasized and will be most evident in the order in which the texts are addressed. Thus, Part I of this volume will discuss the biblical texts that present the primal human traditions in their most pristine form — texts that directly relate information about the figure. Part II will discuss biblical texts that consciously present information about the primal human figure indirectly. This second group of texts presents the same basic figure, but here the primal human is used to form the basis of analogy. That is, the focus is on a protagonist who is likened to the primal human, and not directly on the primal human himself. These instances pose a special problem, since the tenor and vehicle of analogies are often ambiguous. Part III will examine the most vestigial biblical expressions of the primal human conception, in which it has been sublimated, merging with and blending into other images and traditions.

The usefulness of such a study might lie in many areas, but ultimately we might best apprehend it by seeing it through the lens of Clifford Geertz's well-known definition of culture:

> an historically transmitted pattern of meanings embodied in symbols, a system of inherited conceptions expressed in symbolic forms by means

> of which [humans] communicate, perpetuate, and develop their knowledge and attitudes toward life.[36]

The imagery that we are about to explore represents just a part of the system of symbols that comprised ancient Israel's ever-evolving cultural system. The better we can explore aspects of these symbols, the better we are able to forge an understanding of issues and concerns that ancient Israel engaged and which ultimately found expression in the Bible. The idea of the primal human is rife with meaning and presents a most worthy candidate for examination.

Throughout the course of this study, it will become evident that there existed a common conception, manifest in various traditions, of a primordial wise being that functioned as a tool for exploring a host of ideas. Among these are the question of the relation between "humanity" and "deity," and the place of individuals in "elevated" ritual positions of intermediation (e.g., the king, priests, prophets). Authors invoked the symbol of the primal human to engage these questions in various ways. The questions comprise not only the task of discerning the essential boundaries between humanity and the divine, but ultimately the very task of *defining* humanity and the divine.

[36]"Religion as a Cultural System," *The Interpretation of Cultures* (Basic Books, 1973), 89.

Part I

Direct Attestations

Narratives about the Primal Human

1. The Image and Likeness of God

(Genesis 1:26–28)

THE OPENING CHAPTERS OF THE BIBLE preserve the only extant direct attestations of Israelite primal human ideas. In the traditions of Genesis 1–3 when the primal human is discussed, he is the focal point — not a foil or tool of illustration — unlike what we will find in other passages. We will begin here by examining the primal human imagery of Gen 1:26–28, set within the so-called P creation narrative of Gen 1:1–2:4a.

The Text: Genesis 1:26–28

26 וַיֹּאמֶר אֱלֹהִים נַעֲשֶׂה אָדָם בְּצַלְמֵנוּ[a] כִּדְמוּתֵנוּ[b] וּבְעוֹף הַשָּׁמַיִם וּבַבְּהֵמָה [c]וּבְכָל־הָאָרֶץ[c] וּבְכָל־הָרֶמֶשׂ הָרֹמֵשׂ עַל־הָאָרֶץ׃
27 [a]וַיִּבְרָא אֱלֹהִים ׀ אֶת־הָאָדָם [b]בְּצַלְמוֹ
בְּצֶלֶם אֱלֹהִים בָּרָא אֹתוֹ
זָכָר וּנְקֵבָה בָּרָא אֹתָם׃
28 [a]וַיְבָרֶךְ אֹתָם אֱלֹהִים וַיֹּאמֶר[a] לָהֶם אֱלֹהִים פְּרוּ וּרְבוּ וּמִלְאוּ אֶת־הָאָרֶץ וְכִבְשֻׁהָ וּרְדוּ בִּדְגַת הַיָּם וּבְעוֹף הַשָּׁמַיִם [b] וּבְכָל־חַיָּה הָרֹמֶשֶׂת עַל־הָאָרֶץ׃

26 God said, "Let us make humanity in our image, according to our likeness; and let them rule over the fish of the sea, and over the birds of the skies, and over the cattle, and over all the earth, and over every creeping thing that creeps upon the earth."

27 So God created humanity in his image:
in the image of God he created him;
male and female he created them.

28 God blessed them, and God said to them, "Be fruitful and multiply and fill the earth and subdue it. Rule over the fish of the sea and over the birds of the skies and over every living thing that creeps upon the earth."

Notes:

26 [a] Sam וכדמותנו cf. LXX, Vulg; MT is grammatically acceptable as an example of apposition. [b] Some have taken *wāw* followed by Impf as signifying purpose: 'in order that they may rule' (cf. GKC §109; Lambdin §107). [c–c] ובכל הארץ MT, Sam; LXX (και πασης της γης); Syr reflects ובכל חית הארץ. The phrase חית הארץ appears earlier in v. 25 and the variant חיתו ארץ appears earlier in v. 24. The references to בהמה 'cattle' and רמש 'creeping things' may be instructive. In vv. 24–25 both items are placed together in the sequence בהמה...רמש. In v. 24, בהמה...רמש is followed by חיתו ארץ, and in v. 25 חית הארץ precedes בהמה...רמש. Because the reference to חיה appears at the end of v. 24 and at the beginning of v. 25, each case is best taken as a summary statement referring to *all* animals. Some commentators understand חיה as 'wild animals' (e.g., NRSV; Westermann, *Genesis*, 77). The text here, however, makes no distinction between wild and domesticated animals. Thus, rather than a scribal omission, Syr reflects a harmonizing plus. MT, Sam ובכל הארץ is more difficult. It seems out of place and may have been influenced by the presence of על הארץ at the end of v. 26 (for further discussion see R. Hendel, *Genesis 1–11: Textual Studies and Critical Edition* [New York: Oxford University Press, 1998], 122).

27 [a] The initial ויברא followed by the two uses of the Perf ברא suggests that the two clauses with ברא explain the clause with ויברא. The repetition of the Perf ברא followed by the object marker (with m.s. and pl. pronominal suffixes, respectively) suggests that the two statements are in parallel and as such should be construed as a verse, this despite the fact that the editors of BHS have elected not to do so. [b] MT, Sam; בצלמו is lacking in LXX perhaps because of haplography influenced by the identical initial sequence of the first three characters of בצלמו and בצלם.

28 [a–a] LXX reads simply λεγων (= לאמר) perhaps to harmonize with the command to sea creatures and birds in v. 22, where MT, 4QGen[b#d], Sam, and LXX all reflect ויברך אלהים אתם לאמר. Note that in v. 22, Syr reflects the reading ויאמר להם for לאמר, perhaps harmonizing with the textual tradition reflected in LXX v. 28. [b] Syr adds ובבהמה, an apparent harmonization with v. 26 השמים ובבהמה. [c–c] (MT, Sam) Sam החיה for חיה; LXX reads και παντων των κτηνων και πασης της γης και παντων των ερπετων των ερποντων reflecting a harmonization with v. 26 (= ובבהמה ובכל הארץ ובכל הרמש הרמש). Such a harmonization is understandable, given the preceding phrase ובעוף השמים.

This tradition of the creation of the human is set within the context of an account of creation and humanity from the beginning of the cosmos through the demise of the Israelite kingdom. Despite this present use, however, the tradition itself possesses its own oral and perhaps literary prehistory, clearly demonstrated by its connections with Mesopotamian traditions such as Enuma Elish.[37] When isolated from its present literary context of the Pentateuch, the repetitive nature of Gen 1:1–2:4a suggests a liturgy, for which it may, in fact, have been used at some point. Yet, in its present context it is anything but a liturgy. It is presented as a movement of *history* — not in our modern sense, of course, but as history nonetheless, that is, in this passage — in its present context — the creation of the human is presented simply as an event in the past.[38] As such, it forms the beginning of the so-called "primeval history" — the early history of humanity presented in Genesis 1–11. In Genesis 12, this early history of humanity becomes, more specifically, the national history of ancient Israel. The significance of the primeval history within the larger context is primarily that of *aetiology*. Its purpose is to explain where humans "came from" and to say something of what they are, that is, their nature. It also serves as a prelude to the national history of Israel.

But the roots of this passage lie far deeper than in simple aetiology and prelude. To explore these roots is to broaden our understand-

---

[37] Note likewise the striking comparisons between the flood narrative (Genesis 6–9) and the Mesopotamian flood traditions, such as *Enūma Ilu Awīlu*.

[38] Within the past two decades, considerable discussion has focused on the nature of the material in the Pentateuch, specifically with regard to the question of "genre." F. M. Cross identified this material as "epic" (*Canaanite Myth and Hebrew Epic* [Cambridge, Mass.: Harvard University Press, 1973], 6); J. Van Seters and R. N. Whybray compared it with Greek historiographical writings, such as those produced by Herodotus. Van Seters (*In Search of History: Historiography in the Ancient World and the Origins of Biblical History* [New Haven: Yale University Press, 1983], 229–32) construed the Pentateuch as an historical work on a par with Joshua–Kings — the Deuteronomistic History (Dtr) — first identified in the modern period by Noth. Whybray, like Noth on Dtr, suggested that single "author," who drew upon sources to compose the work composed the Pentateuch (R. N. Whybray, *The Making of the Pentateuch: A Methodological Study* [Sheffield: JSOT Press, 1987]). A more recent treatment is that of Friedman, who attempts to trace back, and ultimately attribute, material in Dtr to J. All of this highlights serious interest in the nature of history writing and the extent to which such writing characterizes (if at all) the material in the Pentateuch. For a discussion of the debate over the historical nature of the text, see J. Blenkinsopp, *The Pentateuch: An Introduction to the First Five Books of the Bible* (New York: Doubleday, 1992), 37–42. An outstanding treatment of the many problems inherent in distinguishing genres such as "myth" and "history," is B. Halpern, *The First Historians: The Hebrew Bible and History* (San Francisco: Harper and Row, 1988), especially 266–80.

ing of the passage—its ideas and the society that inherited and shaped them. We may accomplish this by examining the language of Gen 1:26–28 along with other material in Genesis that gives insight into the priestly conception of the primal human. There are two things we may assert about the language used: (1) it derives from the royal imagery of the ancient Near East; and (2) it includes all of humanity. After examining these two observations, we will pursue one additional aspect of the priestly conception of the primal human, the so-called Sethite genealogy. My aim here is to demonstrate that, even in the present context, P recognizes the primal human not only in simple terms as the chronological first human, but, more importantly, as the link between deity and humanity.

The Primal Human and Ancient Near Eastern Royal Imagery

One of the more vexing problems to scholarship has been the question of what is meant by the creation of humanity in the image (*ṣelem*) and likeness (*dəmût*) of God. Several positions can be outlined to illustrate the debate.[39] Many commentators have seen a basic distinction between the two terms *ṣelem* and *dəmût* and, accordingly, have understood them to refer to two separate aspects of humanity, namely a "natural" aspect and "supernatural" aspect. The distinction dates back to the early church and continues predominantly in Roman Catholic theological circles, although in a mutated form.[40] An equally ancient tradition has understood the resemblance to God as a reference to humanity's "spiritual qualities or capacities." Philo of Alexandria interpreted it to refer to the human mental capacity alone.[41]

---

[39]See C. Westermann, *Genesis 1–11: A Commentary* (Minneapolis: Augsburg, 1976), 148–55. See also more recently P. A. Bird, "Male and Female He Created Them: Gen 1:27b in the Context of the Priestly Account of Creation," *HTR* 74 (1981): 129–59.

[40]For brief discussion with references, see Westermann, *Genesis 1–11*, 148–49.

[41]J. R. Levison, *Portraits of Adam in Early Judaism: From Sirach to 2 Baruch* (Sheffield: Sheffield Academic Press, 1988), 66. Levison follows T. Tobin in attributing Philo's interpretation of the image as the human mind to the influence of Platonic thought (see Levison, 66–67; 208 note 21; T. H. Tobin, *The Creation of Man: Philo and the History of Interpretation* [The Catholic Biblical Quarterly Monograph Series; Washington, D. C.: The Catholic Biblical Association of America, 1983], 44–47.

Augustine saw it embodied in the power of the soul, the memory, intellect, and will. This is perhaps the most popular explanation in modern times.[42]

The biblical uses of the terms *ṣelem* and *dəmût* offer limited help in determining their use in Gen 1:26–28. The noun *ṣelem* 'image' is the more easily understood of the two terms: it refers to an image, a physical representation of a thing. The root is generally conjectured to be *ṣlm,* although no verbal form appears in biblical, inscriptional, or later Hebrew.[43] The basic meaning of the noun, however, is quite clear. It refers to statues or idols (Num 33:52; Ezek 7:20; 2 Kgs 11:18); pictorial representations (Ezek 23:14); and physical replicas of various items (tumors and mice, 1 Sam 6:5,11; the male figure, Ezek 16:17).[44] The noun *dəmût*, 'likeness', 'form', or 'appearance', derived from the verb *dāmâ* 'to be like', occurs 25 times in the Hebrew Bible, most frequently in the book of Ezekiel. Because of the close semantic range of the words "likeness," "form," and "appearance," it is often difficult to distinguish definitively among such choices. *HALOT* (1:226) gives 'model' as the primary definition, based on the proposed Arabic cognate *dumyat* 'shape', 'statue' (cf. Tigre *dumat* 'indistinct outline of a figure or an object'). BDB (198) interprets *dəmût* as 'pattern' in 2 Kgs 16:10, where it may to refer to a physical model: 'King Ahaz sent to Uriah the priest a model (*dəmût*) of the altar and its plan (*tabnîtô*) for all of its construction (*ləkol-maᶜăsēhû*)'. In the biblical corpus, nouns used in conjunction with *dəmût* tend to be corporeal objects (face, Ezek 1:10; creatures, Ezek 1:5; human being, Ezek 1:5; throne, Ezek 1:26; Babylonians, Ezek 23:15). One example of a somewhat more abstract noun associated with *dəmût* is the 'glory' of Yahweh (Ezek 1:28); this

---

[42]See Westermann, *Genesis 1–11*, 149.

[43] BDB (853) suggests Arabic *ṣalama* 'to cut off' and thus interprets the essence of *ṣelem* as 'something cut out' on analogy with *pesel* 'idol, image', which derives from the Hebrew root *psl* 'to hew, hew into shape'. The verb *pāsal* is used with reference to hewing stone tablets, building stones, and idols. *HALOT* (3:1028) also cites Palestinian Jewish Aramaic and Syriac 'to provide with sculpture'.

[44] Ps 39:7 and the textually uncertain Ps 73:20 are both examples that *HALOT* (3:1029) rightly attributes to the more prevalent Semitic root *θ̣lm* 'to become dark' (cf. Akk. *ṣalāmu*).

of course is within the context of and summarizes the other examples from Ezekiel 1 just cited.[45] The obvious problem in determining the sense of *ṣelem* and *dəmût* in Gen 1:26–28 is the inherent circularity of defining the human being by means of a reference to *ʾĕlōhîm,* who is already understood in anthropomorphic terms.

Various ancient Near Eastern texts make it quite clear that the language of image and likeness is borrowed from the realm of royal ideology and that the basis for the likeness is *physical,* not spiritual or mental or ethical. In the last century, some began to interpret the phrase as a reference to the external form of the human. H. Gunkel, for example, thought humanity's likeness to God was in terms of appearance or corporeal form.[46]

The interpretation of *ṣelem* and *dəmût* as referring to physical likeness has been strengthened by the ninth-century Aramaic text from Tell el-Fakhariyeh. Both Aramaic cognate words (*ṣlm* and *dmwt*) occur in the text, and both make reference, without doubt, to a statue.

12 *...ṣlm : hdysᶜy*
13 *mlk : gwzn : wzy : skn : wzy : ʾzrn : lʾrmwddt : krsʾh*
14 *wlmʾrk : ḥywh : wlmᶜn : ʾmrt : pmh : ʾl : ʾlhn : wʾl : ʾnšn*
15 *tyṭb : dmwtʾ : zʾt : ᶜbd : ʾl : zy : qdm : hwtr : qdm : hdd*
16 *ysb : skn : mrʾ : ḥbwr : ṣlmh : šm*[47]

12 Image of Hadd-yisᶜî
13 King of Guzan and of Sikan and of Azran for exalting and [?] his throne and for lengthening his life,
14 and that the word of his mouth be good before the gods and the people.
15 This likeness he made bigger than any before.

---

[45] In later literature similar examples occur (e.g., דמות רוח כבוד 'spirit of glory'; 4QShirShab[f]). See *DCH* 2:449.

[46]He based this conclusion on Gen 5:3, 'When Adam had lived one hundred thirty years, he begat a son in his likeness, according to his image', and on the fact that theophanies are frequently presented anthropomorphically. More recently, this position has been maintained by G. von Rad ("Vom Menschenbild des AT," *Der alte und der neue Mensch* [Münich: A. Lempp, 1942], 5–23), W. Zimmerli, (*Das Menschenbild des AT* [Theologische Existenz Heute; Münich: C. Kaiser, 1949]).

[47]Lines 12–16. For text, see A. Abou-Assaf, P. Bordreuil, and A. R. Millard, *La statue de Tell Fekherye et son inscription bilingue assyro-araméenne* (Etudes Assyriologiques. Vol. 7. Paris: Editions Recherche sur les civilisations, 1982). For *dəmût* see also line one.

16 Before Hadad, who dwells at Sikan, Lord of Habur, his image he has erected.

In simple terms what has been erected is a statue *in the likeness of* the king. The meaning of the *presence* of the statue is a separate issue.

The physical likeness of the king to the gods is expressed in the royal hymn found in the beginning of the thirteenth-century B.C.E. Assyrian Epic of Tukulti Ninurta:

16 *ina* (AŠ) *ši-mat* d*Nu-dím-mud ma-ni it-ti šīr* (UZU) *ilāni* (DINGIR.MEŠ) *mi-na-a-šu*
17 *ina* (AŠ) *purussê* (EŠ.BAR) *bēl mātāti* (EN KUR.KUR) *ina* (AŠ) *ra-a-aṭ šas/turri* (ŠÀ.TÙR) *ilāni* (DINGIR.MEŠ) *ši-pi-ik-šu i-te-eš-ra*
18 *šu-ú-ma ṣa-lam* d*Illil* (BE) *da-ru-ú še-e-mu pi-i nišē* (UKU.MEŠ) *mi-lik māti* (KUR)
20 *ú-šar-bi-šu-ma* d*Illil* (BE) *ki-ma a-bi a-li-di ar-ki mār* (DUMU) *bu-uk-ri-šu*

16 By the fate (determined by) Nudimmud (= Ea) his [the king's] form (*mīnāšu*) is considered the flesh of the gods (*šīr ilāni*).
17 By the decision of the lord of the lands he was successfully cast into/poured through the channel of the womb of the gods.[48]
18 He alone is the eternal image of (*ṣalam*) Enlil, attentive to the voice of the people, to the counsel of the land.
20 Enlil raised him like a natural father, after his first-born son (=Ninurta).[49]

Here, the king is the physical likeness of the gods. This is especially explicit in the reference to his being 'cast'. The phrase *šīr ilāni* 'flesh of the gods' is also applied to Gilgamesh in Tablet IX.

The scorpion-man called to his mate:
"He who has come to us flesh of the gods (*šīr ilāni*) is his body (*zumuršu*)."
The scorpion-man's wife answered him:
"Two-thirds of him is god, and one-third human."[50]

---

[48]The filial imagery of the formation of the king and the divine womb is not insignificant and recalls the Judahite doctrine of the king as son of Yahweh (2 Sam 7:14; Ps 2:7; 89:27–28).

[49]Translation adapted from P. Machinist, "Literature as Politics: The Tukulti-Ninurta Epic and the Bible," *CBQ* 38 (1976) 455–82, especially 456–66.

[50]Translation according to A. George, *The Epic of Gilgamesh: The Babylonian Epic Poem and Other Texts in Akkadian and Sumerian* (New York: Barnes & Noble, 1999), 71, lines 48–51.

But physical similarity should not be taken superficially; its intent is to express some deeper reality. Physical similarity suggests a similarity in nature, content, or function. In the case of Tukulti Ninurta, the text expresses the elevation of the king to the status of divinity. The tradition, however, stops short of according Tukulti Ninurta explicit divine status through the use of the divine determinative.[51] In a study on the meaning of images among the Babylonians, H. Hehn demonstrated that an image was sometimes construed as standing in the place of the god, and that the image itself could even be divinized.[52] In the same way, the king was seen to be an "image" of the god and as such represented the god as viceroy. Since this view was first propounded, many more examples have been adduced from Mesopotamia and Egypt.

H. Wildberger and W. H. Schmidt have gathered considerable evidence to demonstrate this notion of image as representative or viceroy.[53] Wildberger argued that *ṣelem* was not the usual word in Israel for a divine image or statue[54] and that this usage has its closest analog in Mesopotamia. He concluded that Gen 1:26 is based on the same royal language, which describes the Mesopotamian king as the *ṣalmu* or *muššulu* (image, likeness). He found further evidence in the use of the term *rdh* 'to rule' and the parallel of Psalm 8, a passage which within the context of creation says with reference to the human, *tamšîlēhû bəmaᶜăśê yādeykā* 'you have caused him to rule over all the works of your hands'. Schmidt argued that the text in Genesis presumes an understanding of image and likeness that was widely held, and therefore found it best explained as a fixed formula, as found elsewhere. The closest parallels were to be found in Egyptian Dynasty 18 texts

---

[51] See Machinist, "Literature as Politics," 467; cf. H. Frankfort, *Kingship and the Gods* (Chicago: University of Chicago Press, 1978), 295–307.

[52] "Zum Terminus 'Bild Gottes'," in *Festschrift Eduard Sachau*, ed. Gotthold Weil (Berlin: G. Reimer, 1915), 36–52.

[53] H. Wildberger, "Das Abbild Gottes," *TZ* 21 (1965): 245–59, 481–501; see especially 251–55. W. H. Schmidt, *Die Schöpfungsgeschichte der Priesterschrift* (WMANT 17; Neukirchen-Vluyn: Neukirchener Verlag, 1964).

[54] Rather, biblical writers more often use *pesel* or *massēkâ* for a divine image.

according to which pharaoh is the representative of God on earth.[55]

C. Westermann argues that the concern of Gen 1:26–27 is "not with human nature, but with the process of the creation of human beings." He explains that this concern with the "process of creation" indicates that the action should be understood as set within the context of "primeval event" as opposed to "history." Hence, Westermann concludes that the statement "God created man [in his image and likeness]" signifies that "something is being said about the beginning of humanity that is not accessible to our understanding."[56] Westermann is certainly correct in assessing this basic aspect of mythological language; but he has gone too far in arguing against attempting "any further determination" of the significance of image and likeness. In view of the extrabiblical evidence and the theme of the human domination of creation in Gen 1:26,28, and Psalm 8, it is most natural to accept the idea of ruling as an important aspect of image and likeness, and not as incidental.[57]

Royal Imagery, the Primal Human, and All Humanity

Having seen that other ancient Near Eastern literatures use the language of image and likeness in royal contexts, we must then recognize that within its present context in Genesis 1, this "royal" imagery applies to the primal human and to humanity in general. Here the tradent intends the audience to understand that "image" and "likeness" pertain to all humans, and not simply to the first man or couple. The best evidence for this is Gen 9:1–7 (P), which gives the rationale for the divine prohibition against the shedding of human blood: the human — that is, *every* human — is created *in the image of God* (*bəṣelem*

[55]W. H. Schmidt, *Schöpfungsgeschichte*, 137–39. For summary see Westermann, *Genesis 1–11*, 152–53.

[56]Westermann, *Genesis 1–11*, 156.

[57]Mesopotamian and Egyptian evidence have continued to play a significant role in the debate since Westermann's comments. See, for example, P. A. Bird, who finds the language of Gen 1:26–27 related to Egyptian and Mesopotamian conceptions, but not based on them. She, instead, seeks the foundation in an unattested Canaanite tradition. P. A. Bird, "Male and Female He Created Them," 142.

ʾĕlōhîm). This understanding of the "image" in Gen 9:1–7 is also expressed within the context of "ruling," although in slightly different terms.

> [1]וַיְבָרֶךְ אֱלֹהִים אֶת־נֹחַ וְאֶת־בָּנָיו וַיֹּאמֶר לָהֶפְרוּ וּרְבוּ וּמִלְאוּ אֶת־הָאָרֶץ׃
> [2]וּמוֹרַאֲכֶם וְחִתְּכֶם יִהְיֶה עַל כָּל־חַיַּת הָאָרֶץ וְעַל כָּל־עוֹף הַשָּׁמָיִם בְּכֹל
> אֲשֶׁר תִּרְמֹשׂ הָאֲדָמָה וּבְכָל־דְּגֵי ם הַיָּם בְּיֶדְכֶם נִתָּנוּ׃

> [1]God blessed Noah and his sons, and said to them "be fruitful and multiply and fill the earth. [2]The fear of you and the dread of you shall be upon every living creature upon the earth, and upon every bird of the air, on everything that creeps on the ground, and on all the fish of the sea; into your hand they are delivered."

The command in Gen 1:28, 'Be fruitful and multiply and fill the earth', continues with 'and subdue (*kbš*) it, and rule (*rdh*) [the animal kingdom]'. Similarly, the same command in Gen 9:1 ('Be fruitful and multiply') continues in v. 2 with 'into your hand they all (i.e., the animal kingdom) are delivered' (*bəyedkem nittānû*). The phrase *nātan bəyād* routinely expresses the idea of domination, political and otherwise (e.g., Deut 1:27; 1 Sam 14:37; 17:47). Psalm 8 presents a similar understanding of humanity in general:

> 5 מָה־אֱנוֹשׁ כִּי־תִזְכְּרֶנּוּ
> וּבֶן־אָדָם כִּי תִפְקְדֶנּוּ׃
> 6 וַתְּחַסְּרֵהוּ מְּעַט מֵאֱלֹהִים
> וְכָבוֹד וְהָדָר תְּעַטְּרֵהוּ׃

> 5 What are humans that you remember them?
> Mortals that you care for them?
> 6 You have made them nearly gods,
> Crowned them with glory and honor.

The figure *ʾĕnôš*/*ben ʾādām* is 'crowned' with 'glory and honor' and rules over creation. The psalmist writes in 8:7, 'you have given them rule (*tamšîlēhû*) over the works of your hands; all things you have placed under their feet'. It may be that this application of royal language to all humans in Genesis and Psalm 8 simply reflects a variant tradition in the ancient Near East, rather than an extension of an earlier and purely royal tradition. In the Neo-Babylonian text VAT 17019, the creation of a first royal figure is clearly an act separate from the creation of common humanity. The text describes the creation of common humanity, which in turn is followed by instructions for the creation of the king:

30 ^{d}*é-a* KA-*šú* DÙ-*ma* DU$_{11}$.GA *ana* ^{d}GAŠAN-AN INIM MU-*ár*
31 ^{d}*Be-let* DINGIR.MEŠ NIN DINGIR GAL.MEŠ *at-ti-ma*
32 *at-ti-ma tab-ni-ma* LÚ.ULÙ^{lu}-*a a-me-lu*
33 *pi-it-*⌜*qí*⌝*-ma* LUGAL *ma-li-*⌜*ku*⌝ *a-me-lu*
34 *ṭa-a-bi u*[*b*]*-bi-ḫi gi-*[*mi*]*r la-a-ni-šú*
35 *ṣu-ub-bi-i zi-i-mi-šú bu-un-ni-i zu-mur-šú*
36 ^{d}*Be-let*-DINGIR.MEŠ *ip-ta-tiq* LUGAL *ma-li-ku* LÚ

30 Ea began to speak, he directed his word to Belet-ili,
31 "Belet-ili, Mistress of the great gods are you.
32 *You have created the common people (lullâ amēlu),*
33 *now construct the king,* the counselor-man *(šarru māliku amēlu)*
34 With goodness envelop his entire form *(gimir lānīšu).*
35 Form his features *(zi-i-mi-šu)* harmoniously; make his body *(zumuršu)* beautiful!"
36 Thus did Belet-ili construct the king, the counselor-man *(šarru māliku amēlu).*[58]

The description of the creation of the king here is reminiscent of the description of the "creation" of Tukulti Ninurta (discussed earlier in this chapter), which placed emphasis on his physical form and referred to him as the 'image of' (*ṣalam*) the god Enlil. The tradition of a separately created royal prototype in Mesopotamia corroborates the idea that there was a royal aspect or motif within Israelite creation traditions, and that such a motif is now variously manifested in the allusions of Gen 1:26–28, Psalm 8, and Ezek 28:11–19.[59] In other words, it is not likely that the royal allusions in these biblical passages reflect either innovation or afterthought. Rather, the imagery may have existed in some form in the *prehistory* of these passages.

---

[58]For text see W. R. Mayer, "Ein Mythos von der Erschaffung des Menschen und des Königs," *Or* 56 (1987): 55–68. Italics in the English translation indicate my emphasis.

[59]See discussion of Ezek 28:11–19 below in chapter 3.

The Primal Human as Exalted Being

Despite the questions that remain concerning the primal human in Gen 1:26–27, it is clear that its description of the creation of *ʾādām* makes lofty assertions about this first human being. The next question we must ask concerns the difference between the first human(s) and their progeny. The P tradition preserves an important idea: it is clear that Adam and Eve are not on the same level as their creator, yet it is equally clear that they are not quite the same as those who follow and eventually populate the earth. There is an obvious and understandable awareness that Adam stands between God and humanity. This awareness finds expression in the so-called Sethite genealogy of Genesis 5. The Sethite genealogy was incorporated by P from an ancient document known as the *sēper tôlədôt ʾādām,* 'the document of the generations of Adam'.[60] The genealogy begins in 5:1 with a restatement of the creation of the human pair first recorded in 1:26–27:

> [1]זֶה סֵפֶר תּוֹלְדֹת אָדָם בְּיוֹם בְּרֹא אֱלֹהִים אָדָם בִּדְמוּת אֱלֹהִים עָשָׂה
> אֹתוֹ׃ [2]זָכָר וּנְקֵבָה בְּרָאָם וַיְבָרֶךְ אֹתָם וַיִּקְרָא אֶת־שְׁמָם אָדָם בְּיוֹם
> הִבָּרְאָם׃ [3]וַיְחִי אָדָם שְׁלֹשִׁים וּמְאַת שָׁנָה וַיּוֹלֶד בִּדְמוּתוֹ כְּצַלְמוֹ וַיִּקְרָא
> אֶת־שְׁמוֹ שֵׁת׃
>
> [1]This is the document of the generations of Adam. When God created Adam, in the likeness of God he made him. [2]Male and female he created them. He blessed them and named them *ʾādām* when he created them. [3]Now Adam was nine hundred thirty years old. He begat in his image and according to his likeness and named him Seth.

That *bəṣelem ʾĕlōhîm bārāʾ ʾōtô* of Gen 1:27 is reflected here in the variant form *bidmûṯ ĕlōhîm (ʿāśâ ʾōtô)* suggests that it may have been a common aphorism. Regardless, the transition from *ʾādām* 'humanity' to Adam, the primal human figure, is seamless, the play on words unmistakable. It is precisely the double entendre in the text here that reveals layers of meaning in discussing the primal human figure.

Here, just as God creates (*bārāʾ*) humanity in his (image and)

---

[60] The distinct style of the genealogy is clear: '[These are the generations of] N_1. N_1 was X years old and begot N_2. The days of N_1 after he begot N_2 were Y years, and he begot sons and daughters. [All the days of the life of N_1 were Z years and he died]' (F. M. Cross, *Canaanite Myth and Hebrew Epic,* 301).

likeness, so also does Adam beget Seth in his image and likeness. This description expresses the first man's continuity with and disjuncture from God as well as his continuity with and disjuncture from those who follow him. The statement, at least in part, is certainly a commentary on the generational transmission of physical characteristics, yet it is curious that it does not recur for *any* of the subsequent generations. The highly formulaic and repetitive nature of the genealogy makes it difficult to explain this omission as an attempt to avoid redundancy. Rather, a more likely explanation is that the structure conveys meaningful information about Adam himself, and the generation that he and Eve represent. They are the only generation created by God without the agency of a woman. From the standpoint of genealogical structure, to place God at the beginning of the genealogy, as the text has done, is to place him in the position of "father" to Adam. That is, a genealogical relationship between God and the first man is suggested in the structure of Gen 5:1–3, even though 5:1 does not use the verb 'beget' with reference to God. Like Gen 1:27, Gen 5:1–3 does not use the imagery of "siring" offspring through the agency of another (Hiphil form of the root *yld* 'to bear', thus 'to cause to bear'), but that of creating (*bārāʾ*) a being, by himself. (One can not conclude, however, that similar traditions that draw upon birth imagery did not exist in ancient Israel. Explicit "birth" language used with regard to the first human's origins appears in Job 15:7, where Eliphaz facetiously asks Job, 'were you *born* the first human? Were you *brought forth* before the hills?'[61]).

The special nature of such an intermediate position was recognized not only here in the text of Genesis 5; it was widely acknowledged throughout the ancient Near East. This is particularly clear in genealogical and dynastic succession contexts. In the Egyptian royal ideology of Manetho and the Turin Canon, a dynastic succession is posited that begins with the gods and ends with known historical kings. According to both traditions, kingship begins with the gods of the Ennead. These are succeeded by demigods and 'dead ones' (νεκυες, *mānēs*). Menes of Dynasty I follows and heads the list of hu-

[61] See the discussion of Job 15 in chapter 4.

man rulers. In the Turin Canon, Seth is followed by a 'Horus of the Gods', which later recurs before a group called the 'Followers of Horus', who are then followed by Menes.[62] Most Egyptologists consider the traditionally listed predynastic kings of Egypt nonhistorical demigods.[63] What is interesting here for our discussion is the movement from divine to human, passing through an intermediate phase.

Isa 19:11, a passage alluded to earlier, reveals an awareness in Israel of this same type of tradition. The prophet mocks the true ignorance of the 'princes' of Zoan and the 'wise counselors' of Pharaoh:

אַךְ־אֱוִלִים שָׂרֵי צֹעַן
חַכְמֵי יֹעֲצֵי פַרְעֹה עֵצָה נִבְעָרָה
אֵיךְ תֹּאמְרוּ אֶל־פַּרְעֹה בֶּן־חֲכָמִים
אֲנִי בֶּן־מַלְכֵי־קֶדֶם׃

> Surely, the princes of Zoan are fools
> the "wise" counselors of Pharaoh give stupid counsel
> How can you say to Pharaoh, "I am one of the sages,
> a descendant of ancient kings"?

Unfortunately, the prophet doesn't elaborate further on this connection to the past. It remains clear, nonetheless, that according to the prophet's understanding, the basis of the Egyptians' claims to wisdom lay in a perceived connection with ancestors of earliest antiquity. The appeal is explicitly genealogical.

A similar intermediate position for the first humans may be seen in the writings of the Phoenician Sakkunyaton.[64] Fragments of Sakkun-

---

[62]See A. Gardiner, *Egypt of the Pharaohs* (Oxford: Oxford University Press, 1961), 421.

[63]See B. G. Trigger, et al., *Ancient Egypt: A Social History* (Cambridge: Cambridge University Press, 1983), 44.

[64]Sakkunyaton's dates are disputed, ranging from the second millennium B.C.E. (O. Eissfeldt, "Religionsdokument und Religionspoesie, Religionstheorie und Religionshistorie: Ras Schamra und Sanchunjaton, Philo Byblius und Eusebius von Cäsarea," *Kleine Schriften* 2 [1963], 130–44; "Zur Frage nach dem Alter der phönizischen Geschichte des Sanchunjaton," *Kleine Schriften* 2 [1963], 127–29) to the mid-first millennium B.C.E. (W. F. Albright, "Recent Progress in North-Canaanite Research," *BASOR* 70 [1938]: 24 ; *From the Stone Age to Christianity*, [Baltimore: Johns Hopkins University Press, 1957], 230; *Archaeology and the Religion of Israel*, [Baltimore: Johns Hopkins University Press, 1956], 70; F. Løkkegaard, "Some Comments on the

yaton's work were preserved in the *Phoenician History* of Philo of Byblos (late first century C.E.), parts of which, in turn, were included by the church father Eusebius (fourth century) in his *Praeparatio evangelica*. In one fragment Sakkunyaton recounts the history of culture. It is essentially a genealogical text, which includes both cosmogony and the advent of culture:

> [7] εἶτά φησιν γεγενῆσθαι ἐκ τοῦ Κολπία ἀνέμου καὶ γυναικὸς Βάαυ (τοῦτο δὲ νύκτα ἑρμηνεύει) Αἰῶνα καὶ πρωτόγονον, θνητοὺς ἄνδρας, οὕτω καλουμένους εὑρεῖν δὲ τὸν Αἰῶνα τὴν ἀπὸ δένδρων τροφήν. ἐκ τούτων τοὺς γενομένους κληθῆναι Γένος καὶ Γενεάν, καὶ οἰκῆσαι τὴν φοινίκην αὐχμῶν δὲ γεομένων τὰς χεῖρας εἰς οὐρανὸν ὀρέγειν πρὸς τὸν ἥλιον "τοῦτον γάρ," φησί "θεὸν ἐνόμιζον μόνον οὐρανοῦ κύριον, Βεελσάμην κλοῦντες, ὅ ἐστι παρὰ φοίνιξι κύριος οὐρανοῦ, Ζεὺς δὲ παρ' Ἕλλησιν....
> [9] 'Εξῆς φησιν ἀπὸ Γένους Αἰῶνος καὶ Πρωτογόνου γεννηθῆναι αὖθις παῖδας θνητούς, οἷς εἶναι ὀνόματα φῶς καὶ Πῦρ καὶ ˆ όξ. "οὗτοι," φησίν, "εὗρον ἐκ παρατριβῆς ζύλων πῦρ καὶ τὴν χρῆσιν ἐδίδαξαν. υἱοὺς δὲ ἐγέννησαν οὗτοι μεγέθει τε καὶ ὑπεροχῇ κρείσσονας, ὧν τὰ ονοματα τοις ορεσιν επετεθη ων εκρατησαν
>
> [7] Then he says that from the wind Colpia [Κολπία ἀνέμου] and his wife Baau [γυναικὸς Βάαυ] (this means night) were born Aeon and Protogonos, mortal men [θνητοὺς ἄνδρας], so called. Aeon discovered the nourishment from trees. Their offspring were called Genos and Genea and they settled Phoenicia. When droughts occurred, they raised their hands to heaven, toward the sun. "For," he says, "they considered him, the lord of heaven, to be the only god and called him Beelsamen, which is 'Lord of Heaven' in Phoenician, Zeus in Greek....
> [9] Next he says that from Genos, son of Aeon and Protogonos, there were born further mortal children [παῖδας θνητούς], whose names are Light, Fire, and Flame. He says, "These discovered fire by rubbing sticks of wood together, and they taught its usefulness. They begot sons greater in size and stature, whose names were given to the mountains over which they ruled."[65]

Sanchuniathon Tradition." *ST* 8 [1955]: 51–76) to the Persian or Hellenistic Period (M. L. West, ed., *Hesiod: Theogony* [Oxford: Clarendon, 1966], 26). For further discussion, see H. Attridge and R. Oden, *Philo of Byblos: The Phoenician History* (The Catholic Biblical Quarterly Monograph Series; Washington, D. C.: The Catholic Biblical Quarterly Association of America, 1981), 1–9; A. Baumgarten, *The* Phoenician History *of Philo of Byblos: A Commentary* (Leiden: E. J. Brill, 1981).

[65]Eusebius, PE 1.10.6–14; text and translation from Attridge and Oden, *Philo of Byblos*, 41–42.

Here, the earliest humans, called 'mortal men' (θνητοὺς ἄνδρας), are viewed as the offspring of the divine cosmic elements. They are also viewed as "culture bearers," reminiscent of the earliest humans recounted in the Cainite genealogy of Gen 4:17–22. The propinquity of early humanity to the divine is also made clear in texts such as Gen 6:1–3, according to which divine beings freely intermarry with human beings and produce offspring.

In several Mesopotamian creation traditions, such as Enuma Elish and the Atrahasis Epic, the gods construct the first human beings from the blood of a sacrificed god or from a slaughtered deity's flesh and blood mixed with clay.[66] Whether the scribes and audience understood this to mean that a divine element was present in *all* of humanity is not made explicit, although it is suggested in the bilingual Sumerian and Babylonian text *KAR* 4, the copy of which dates to ca. 800 B.C.E. The text (obverse) also reflects the tradition in which the gods create the first humans from the blood of two slain gods.[67]

21 *ilū*^{meš} *rabûtu*^{meš} *šu-ut iz-zi-zu*
22 ^{d}*A-nun-na-ku mu-ši-im ši-ma-ti*
23 *ki-la-lu-šu-nu* ^{d}*en-líl ip-pa-[lu-šu]*
24 *i-na uzu-mú-a*^{ki} *ri-ki-is šamê u erṣeti*^{ti}
25 ^{d}*lamga* ^{d}*lamga ini-iṭ-bu-ḫa*
26 *i-na da-me-šu-nu ini-ib-na-a a-mi-lu-ta*

21 The great gods who were present
22 The Anunnaki, who fix the destinies
23 Both of them, made answer to Enlil:
24 "In Uzumua, the bond of heaven and earth,
25 Let us slay (two) Lamga gods.
26 With their blood let us create humankind.

Later, the text (reverse) apparently names these first two humans:

52 ^{d}*ul-le-gar-ra an-né-gar-ra*
53 *šu-me-šu-nu ta-za-na-kà[r]*

---

66 Enuma Elish, tablet VI, lines 31–33, in the edition of B. R. Foster, *Before the Muses* (2 vols.; Bethesda, Md.: CDL Press, 1993), 1.385; Atrahasis (OB version) lines 190–212, idem, 1.165.

67 For text and translation, see G. Pettinato, *Das altorientalische Menschenbild und die sumerischen und akkadischen Schöpfungsmythen* (Heidelberg: Carl Winter Universitätsverlag, 1971), 75–78.

> 52 dUllegarra and Annegarra
> 53 You will call their names...

The scribe here has placed the divine determinative before the two names (*dul-le-gar-ra an-né-gar-ra*).[68] The fact that the divine determinative normally distinguishes *divine* or *semidivine* beings from "stock" *human* beings (indicated by the absence of the divine determinative) suggests that these two figures were considered qualitatively different from their descendants.

To return to the motif of the culture bearer — which we shall again revisit in the next chapter — the succession from divine to human and the significance placed on intermediate status are likewise evident in the Mesopotamian tradition of the *apkallu*, expressed in texts such as "the Etiological Myth of the Seven Sages" (LKA 76), "the Uruk List" (W20 030, 7), and the *bīt mēseri* text.[69] In the "Myth of the Seven Sages" and the *bīt mēseri* text, the earliest of these *apkallu* figures are called 'the pure *purādu*-fish of the sea', those who were 'formed in the river, [and] who guide the plans of the heaven and the earth'.[70] The texts continue by listing the succession of later *apkallu*s, who differ from the earlier by being of *human origin* (*ilitti amēlūti*).[71] S. Denning-Bolle appropriately summarizes what is expressed in these three texts:

> Hence, we see an interesting development here. First, the *apkallu*s were created by the gods [and] were bearers of knowledge and crafts. Second, these first *apkallu*'s were succeeded by *apkallu*'s of human origin. Finally, these in turn were followed by the *ummānu*'s, the 'scholars' (UVB 18:44–45). The *ummānu*'s were not only craftsmen in the

---

[68]Following Pettinato's reading (ibid., 76). The proper reading and interpretation of these names is still unsettled (ibid., 81 note on lines 51–53). Note also the older convention, which read the first two signs of the second name *dzal* (for *an-né*); see for example A. Heidel, *The Babylonian Genesis* (Chicago and London: University of Chicago Press, 1951), 69.

[69]See E. Reiner, "The Etiological Myth of the Seven Sages," *Or* 30 (1961): 1–11, J. J. A. van Dijk, UVB 18 (1962): 45, and R. Borger, "Die Beschwörungsserie Bīt Mēseri und die Himmelfahrt Henochs," *JNES* 33 (1974): 183–96, respectively.

[70]See Reiner, "Etiological Myth," 2; Borger, "Bīt Mēseri," 92.

[71] See Reiner, "Etiological Myth," 3.

> traditional sense but were leaders of the schools as well. A line extends, then, from the initial divinely created *apkallu*'s to the *ummānu*'s who were responsible for human intellectual development.[72]

Throughout the ancient Near East, there are traditions which address the intersection of human and divine realms in earliest times, and as we have seen, the priestly tradent expresses interest in this intersection as well. This, again, is apparent in the structure of the Sethite Genealogy in Genesis 5, which may be summarized as follows: the list of Adam and his descendants is begun *with God;* similar language (involving the idea of "image and likeness") is used to describe the acts of God and of Adam in producing the next person listed; this language ceases with Adam and is not repeated again for any of the subsequent acts of engendering, despite the highly formulaic and repetitive nature of the text.

We have drawn attention to three aspects of the language used about the creation of humanity, and of the primal human, in P: first, that it derives from the royal imagery of the ancient Near East; second, that in this particular context, such royal imagery embraces all of humanity; and finally, that P (especially in the Sethite genealogy of Gen 5:1–3) recognizes the primal human not only in simple terms as the chronological first man, and not simply as a type of royalty, but as the link between deity and humanity.[73]

---

[72]S. Denning-Bolle, *Wisdom in Akkadian Literature: Expression, Instruction, Dialogue* (Leiden: Ex Oriente Lux, 1992), 49–50.

[73]In discussing the culture-bearer tradition in the ancient Near East and its connection to biblical and apocryphal traditions, P. Hanson aptly notes the significance of the location of these beings in primordial times. Figures were placed in the remote past to make a point about their relation to deity: "the primordial age is so distant as to invite confusion between human and divine." This period was generally considered to be one in which "the lines between gods and humans were yet indistinct." P. D. Hanson, "Rebellion in Heaven, Azazel, and Euhemeristic Heroes in 1 Enoch 6–11," *JBL* 96 (1977): 228.

2. In (and out) of the Garden of Eden (Genesis 2–3)

GENESIS 2–3 PROVIDES THE SECOND and final direct glimpse in the canonical record into the ancient Israelite conception of the primal human. In its present context it reads as a complementary account of the creation of humanity, fuller than the summary-like tradition of Gen 1:26–28. Like the material in Genesis 1, it too in its larger framework has what might be construed as a historical orientation. But a closer and comparative reading of the language and imagery of narrative — a reading sensitive to its ancient Near Eastern background — reveals a great deal about the prehistory of the present text and brings us to a greater appreciation of the tradition's rich and vibrant nature.

For reasons given in the Introduction, I should like to examine four elements or *topoi* which naturally stand out as crucial to the shape and direction of the narrative. These topoi may be summarized as *creation, location, wisdom,* and *conflict*. Later, we will see that, in different degrees, these reemerge in and govern other expressions or presentations of the primal human in the Hebrew Bible. Finally, we will examine the most widely accepted — and debated — ancient Near Eastern analogue, the Adapa tradition, seeking to discover how it can contribute to our understanding of Israel's primal human conception as it appears in Genesis and in other biblical texts.

Creation and the Primal Human

The Text: Genesis 2:4b–7

[4b]בְּי֗וֹם עֲשׂ֛וֹת יְהוָ֥ה אֱלֹהִ֖ים [a]אֶ֥רֶץ וְשָׁמָֽיִם[a]׃ [5]וְכֹ֣ל ׀ שִׂ֣יחַ הַשָּׂדֶ֗ה טֶ֚רֶם
יִֽהְיֶ֣ה בָאָ֔רֶץ וְכָל־עֵ֥שֶׂב הַשָּׂדֶ֖ה טֶ֣רֶם יִצְמָ֑ח כִּי֩ לֹ֨א הִמְטִ֜יר יְהוָ֤ה
אֱלֹהִים֙ עַל־הָאָ֔רֶץ וְאָדָ֣ם אַ֔יִן לַעֲבֹ֖ד אֶת־הָֽאֲדָמָֽה׃ [6]וְאֵ֖ד יַעֲלֶ֣ה מִן־הָאָ֑רֶץ
וְהִשְׁקָ֖ה אֶֽת־כָּל־פְּנֵֽי־הָאֲדָמָֽה׃ [7]וַיִּ֩יצֶר֩ יְהוָ֨ה אֱלֹהִ֜ים אֶת־הָֽאָדָ֗ם עָפָר֙
מִן־הָ֣אֲדָמָ֔ה וַיִּפַּ֥ח בְּאַפָּ֖יו נִשְׁמַ֣ת חַיִּ֑ים וַֽיְהִ֥י הָֽאָדָ֖ם לְנֶ֥פֶשׁ חַיָּֽה׃

> [4b]When Yahweh God made the earth and the heavens, [5]when no plant of the field was yet in the earth and no herb of the field had yet sprouted because the Yahweh God had not caused it to rain upon the earth and there was no human to work the ground, [6]a stream would rise from the earth and water the entire face of the ground. [7]Then Yahweh God formed the human out of dust from the earth, breathed into his nostrils the breath of life and the human became a living being.

Notes:

4 [a–a] Sam and Syr read שׁמים וארץ, cf. Gen 1:1.

The primal human is explicitly set within the context of creation. Furthermore, the narrative gives specific information about the creation of the primal human. Both ideas establish a framework for understanding the basic conception of this character. Placing the primal human within the context of creation does not appear to be unusual; in fact, it may seem a bit difficult to conceive of him in any other setting. This contextualization of the primal human allows us to make the observation that the primal human *belongs to the primordium* and in this respect is very much unlike us (or unlike the original Israelite audience). As I have discussed in chapter 1, this context allows the primal human to enjoy a special unmediated relationship with God — it puts the primal human in a different kind of relationship to God from ordinary humans.

The description of the creation of the primal human reveals both the similarities and the differences between the primal human and the rest of humanity. He was created in a fashion unlike any other human being, but he was composed of the same materials.

In the Garden: The Location of the Primal Human

The location of the primal human is of central importance to the narrative and to the conception of the primal human. Genesis 2–3 locates the primal human in the garden of Eden.

The Text: Genesis 2:8–14

> [8]וַיִּטַּע יְהוָה אֱלֹהִים גַּן־בְּעֵדֶן מִקֶּדֶם[a] וַיָּשֶׂם שָׁם אֶת־הָאָדָם אֲשֶׁר יָצָר׃
> [9]וַיַּצְמַח יְהוָה אֱלֹהִים[a] מִן־הָאֲדָמָה כָּל־עֵץ נֶחְמָד לְמַרְאֶה וְטוֹב לְמַאֲכָל
> וְעֵץ הַחַיִּים בְּתוֹךְ הַגָּן וְעֵץ הַדַּעַת טוֹב וָרָע׃ [10]וְנָהָר יֹצֵא מֵעֵדֶן

לְהַשְׁקוֹת אֶת־הַגָּן וּמִשָּׁם יִפָּרֵד וְהָיָה לְאַרְבָּעָה רָאשִׁים׃ [11]שֵׁם הָאֶחָד
פִּישׁוֹן הוּא הַסֹּבֵב אֵת כָּל־אֶרֶץ הַחֲוִילָה[a] אֲשֶׁר־שָׁם הַזָּהָב׃ [12]וּזֲהַב
הָאָרֶץ הַהִוא[a] טוֹב[b] שָׁם הַבְּדֹלַח וְאֶבֶן הַשֹּׁהַם׃ [13]וְשֵׁם־הַנָּהָר הַשֵּׁנִי גִּיחוֹן
הוּא הַסּוֹבֵב אֵת כָּל־אֶרֶץ כּוּשׁ׃ [14]וְשֵׁם[a] הַנָּהָר הַשְּׁלִישִׁי חִדֶּקֶל הוּא
הַהֹלֵךְ קִדְמַת אַשּׁוּר וְהַנָּהָר הָרְבִיעִי הוּא פְרָת׃

[8]Yahweh God planted a garden in Eden in the east. There he placed the human which he had formed. [9]Yahweh God caused to sprout from the ground every tree pleasing in appearance and good for food. The tree of life was in the middle of the garden, as was the tree of the knowledge of good and evil. [10]A river went forth from Eden to water the garden; from there it divides and becomes four head waters. [11]The name of the first is Pishon; it is the one that goes around the entire land of Havilah, where there is gold. [12]And the gold of that land is fine. Bdellium and onyx stone are there. [13]The name of the second river is Gihon; it is the one that goes around the entire land of Cush. [14]The name of the third river is Tigris; it is the one that runs east of Assyria. The fourth river is Euphrates.

Notes:

8 [a] מקדם is often translated 'in the east'. It may, however, also be understood temporally and thus translated 'in the earliest of times' or 'from time immemorial'. Cf. Isa 45:21; 46:10; Ps 74:12; 77:6,12; 143:5.

9 [a] LXX adds ετι = עוד, cf. Gen 1:11–12.

11 [a] Some manuscripts of the Samaritan Pentateuch read חוילה (cf. LXX ευιλατ). It may be that the ה of MT החוילה reflects a dittography based on the graphic confusion of ה and ח.

12 [a] The reading of the Qere is supported by the Sam and Targ (Onq). It is more than likely the result of the graphic confusion of ו and י, as occurs frequently in the Pentateuch. [b] Sam adds מאד, not reflected in MT, Syr, or G.

14 [a] LXX omits ושם.

What does Eden signify and why was the human placed in it? Two suggestions may be made about the J narrative's Eden and the location of the primal human within it. The first is that Eden signifies a locus for divine activity. This idea applies to the narrative in its present context, but undoubtedly hails from a long-standing and widespread notion in the ancient Near East. The second suggestion is that Eden corresponds to and provides an archetype for the temple as the place where human and divine meet. This idea belongs to the prehistory of the tradition, but in the present form of the narrative, it has been subordinated to other concerns of the narrator and has become, in effect, a nonfunctioning vestige. After exploring these two suggestions, we will consider the role of the primal human in that garden setting.

Eden: A Locus for Divine Activity

It may be reasonably demonstrated by analysis of the text and relevant parallels that the garden of Eden was understood to be a locus for divine activity.

An attempt to understand the significance of the garden must begin with what is presented in the text. The text, however, gives few explicit clues regarding this significance. Where the narrative is silent one must attempt to fill the gaps with comparative evidence from other ancient Near Eastern materials. There are sufficient references to gardens in ancient Near Eastern literature to allow us to gain insight into the general conception. It seems clear that ancient Israel shared a common understanding of phenomena that are similarly described elsewhere, although we should be careful not to assume this without proper inquiry. Every narrative relies to a degree on unstated notions. It appears that there existed generally conceived notions of the meaning of the garden, which rendered further elaboration unnecessary. In the desert-conscious world view of the ancient Near East, it is understandable that one conception of a divine habitation or 'seat' was a spring-fed garden oasis.[74]

After introducing the creation of the garden and the placing of the human within it, the text continues with a description of two of its features: trees and the subterranean water source (*nāhār*, called *ʾēd* in v. 6) which watered the garden. As we shall see, comparative ancient

---

[74]The word *ʿēden* has traditionally been traced back to two possible derivations. Many scholars, beginning with F. Delitzsch, have traced the name to the Sumerian word edin, 'steppe, plain', later borrowed by the Babylonians as *edinu*. At some point, the word passed from Akkadian into Hebrew (and perhaps Ugaritic). Both BDB and KB lean toward this Sumerian/Akkadian origin. Others, however, hold that *ʿēden* was, in fact, Semitic in its origin and belongs to a root that conveys the idea 'abundant, lush' (reflected in the LXX translation of v. 13 εν τη τρυφη του παραδεισου); however, Hebrew has been the sole language to preserve the root with this sense. For discussion see A. R. Millard, "The Etymology of Eden," *VT* 34 (1984): 103–5; T. Hiebert, ***The Yahwist's Landscape*** (New York and Oxford: Oxford University Press, 1996), 51–62.

Near Eastern materials reveal that both features are intimately related with the notion of divinity and figure prominently in descriptions of divine habitations and places where deities may be found.

The trees of the garden are described in Gen 2:9. The garden possessed a seemingly exhaustive variety of fruit-bearing trees (*kol-ᶜēṣ neḥmād ləmarʾeh wəṭôb ləmaʾăkāl),* 'every tree that is delightful in appearance and good for food'. The tree of life and the tree of the knowledge of good and evil are also in the garden (v. 9) and are similarly described in 3:6 as *ṭôb...ləmaʾăkāl* 'good for food' and *taʾăwâ...lāᶜênayim* 'pleasing to the eyes' as well as *neḥmād...ləhaśkîl* 'desirable for making one wise'.

Trees and groves figured prominently in the religious literature of ancient Israel as a signifier of divine presence.[75] Trees play a significant role in patriarchal religion, in which cultic sites were said to be erected at locations of unusual holiness, where God had manifested himself. In Hebron Abraham built an altar to Yahweh at 'the oaks of Mamre' (Gen 13:18). The oaks of Mamre also provided the setting for Abraham's encounter with the divine in the form of three mysterious visitors (Gen 18:1–2). In Beersheba, a popular holy site in the patriarchal stories, Abraham planted a tamarisk and 'called there on the name of Yahweh' (Gen 21:33). Abraham is reported to have traveled into Palestine specifically to a place in the vicinity of Shechem called the oak of Moreh ('teacher'; Gen 12:6).[76] Shechem's sacred tree (or grove—it appears in the plural in Deut 11:30) is associated with oracle-giving (or higher knowledge) elsewhere, as it is called the diviners' oak (*ʾēlôn məᶜônənîm;* Judg 9:37). We are reminded that Deborah, the prophet, judged Israel beneath the 'palm of Deborah'. Jacob was believed to have buried the family idols at the oak near Shechem (Gen 35:4). Even beyond the patriarchal narratives, when Joshua performed the covenant renewal ceremony at Shechem, he commemorated the occasion by setting up a monolith 'under the oak in the sanctuary of Yahweh' (Josh 24:26). It appears that this was also the site where, during the period

---

[75]Commentators have long pointed out the existence of sacred trees among the peoples of Arabia. See H. Ringgren, *Israelite Religion* (Philadelphia: Fortress, 1966), 25 and references cited.

[76]Also interpreted as 'oracle-giver' (e.g., Ringgren, ibid.).

of the League, Abimelech was officially made king (Judg 9:6).

The popularity of the notion of sacred trees and groves as the locus of divine activity was so extensive as to pose a threat to the centralized Yahwistic cult. Deuteronomy, known for its insistence upon centralized worship, expresses the understanding of how widespread this notion was in Israel and among its neighbors: 'you shall surely destroy all the places where the nations whom you shall dispossess served their gods, upon the high mountains and upon the hills and under every green tree' (Deut 12:2). This perception is likewise reflected in prophetic polemics. Hosea complains concerning the place where the people are apparently seeking oracles:

עַל־רָאשֵׁי הֶהָרִים יְזַבֵּחוּ וְעַל־הַגְּבָעוֹת יְקַטֵּרוּ
תַּחַת אַלּוֹן וְלִבְנֶה וְאֵלָה כִּי טוֹב צִלָּהּ

Atop the mountains they sacrifice, upon the hills they make offerings, under oak, poplar, and terebinth, for its shade is good. (4:13a, cf. v. 12)

These cultic places came to be popularly described in Deuteronomistic literature as *ᶜal haggəbāᶜôt wətaḥat kol-ᶜēṣ raᶜănān* 'on the hills and under every green tree' (2 Kgs 16:4; cf. 1 Kgs 14:23 'on every high hill [*ᶜal kol-gibᶜâ gəbōhâ*] and under every green tree'; cf. also 2 Kgs 17:10). Particularly significant for our discussion is interest in 'green' trees (Jer 2:20; 3:6; Ezek 6:13; Isa 57:5).[77] This depiction reveals an interest not simply in trees, but in luxuriant trees.[78]

The association of these sacred groves with the concept "garden" is made by Isaiah, who warns those who forsake Yahweh, 'You shall be ashamed of the oaks (*ʾēlônîm)* in which you delighted; and you shall blush for the gardens (*haggannôt*) which you have chosen' (1:29). The garden imagery continues in the statement that follows: 'for you shall be like an oak (*ʾēlâ*) whose leaf withers, and like a garden (*gannâ*) without water' (v. 30). Later, Second Isaiah revealed again the extent to which gardens carried a cultic significance, when he wrote concerning 'a people who provoke me to my face continually, sacrificing in gardens' (*gannôt*; Isa 65:3).

---

[77]Ezekiel (6:13) adds to 'every green tree' 'every leafy oak'.

[78]It is also to be observed that the trees are alive, *not dead*.

The religious significance attached to trees is easily observed in Mesopotamia as well. Sacred trees are a well-known feature in Mesopotamian iconography, and trees figure prominently in descriptions of divine dwellings. In the Gilgamesh Epic the divine dwelling is explicitly described as a mountainous forest of tall luxuriant cedars. The text refers to this as KUR-*ú* GIŠ.ERIN *mu-šab* DINGIR.MEŠ *pa-rak* *dIr-ni-ni*, 'the Mountain of Cedar, seat of gods and Irnini's throne'.[79] The passage continues by emphasizing the height of the trees and their ability to provide superior shade.

> They stood there marvelling at the forest,
> gazing at the lofty cedars.
>
> [On the] face of the mountain the cedar proffered its abundance,
> its shade was sweet and full of delight.[80]

The reference to the trees' ability to provide shade is a feature we have already seen in the words of Hosea (4:13). This description in Gilgamesh also recalls the description of the tree in the garden of God in Ezekiel 31. The tree is described as a magnificent physical specimen:

7 וַיִּיף בְּגָדְלוֹ
בְּאֹרֶךְ דָּלִיּוֹתָיו
כִּי־הָיָה שָׁרְשׁוֹ אֶל־מַיִם רַבִּים׃
8 אֲרָזִים לֹא־עֲמָמֻהוּ בְּגַן־אֱלֹהִים
בְּרוֹשִׁים לֹא דָמוּ אֶל־סְעַפֹּתָיו
וְעַרְמֹנִים לֹא־הָיוּ כְּפֹארֹתָיו
כָּל־עֵץ בְּגַן־אֱלֹהִים לֹא־דָמָה אֵלָיו בְּיָפְיוֹ׃
9 יָפֶה עֲשִׂיתִיו בְּרֹב דָּלִיּוֹתָיו
וַיְקַנְאֻהוּ כָּל־עֲצֵי־עֵדֶן אֲשֶׁר בְּגַן הָאֱלֹהִים׃

---

[79]Gilgamesh V 6; text according to S. Parpola, *The Standard Babylonian Epic of Gilgamesh* (State Archives of Assyria Cuneiform Texts 1. Helsinki: The Neo-Assyrian Text Corpus Project, University of Helsinki Press, 1997), 87; translation adapted from George, *The Epic of Gilgamesh*, 39, line 6. Irnini is most likely another name for Ishtar. The phrase *mūšab ilānī* recalls *môšab ʾĕlōhîm* in Ezek 28:3. See further the discussion of Ezek 28:1–10 in chapter 5.

[80]Gilgamesh V 1–2, 8–10; translation according to George, *The Epic of Gilgamesh*, 39.

> 7 It was beautiful in its greatness,
> *in the length of its branches*
> for all its roots went down to abundant waters.
> 8 The cedars in the garden of God could not rival it,
> nor the fir trees *equal its boughs;*
> the plane trees were as nothing *compared with its branches;*
> no tree in the garden of God was like it in beauty.
> 9 I made it beautiful *in the mass of its branches*
> and all the trees of Eden envied it, that were in the garden of God.

Here, the prophet reveals the commonly understood association of luxuriant trees with the concept "garden." This passage is of obvious interest for our purposes, because of the phrase *gan ʾĕlōhîm* 'garden of God'. For not only do we have another description of a garden distinguished by trees (as is the case in the garden of Eden in Genesis), but this provides further evidence that the garden setting in Genesis was understood to be a locus for divine activity. This combination of ideas has already been alluded to above, found in the words of Isaiah (1:29–30).

The description of the trees in the garden of Eden in Genesis 2:9 depicts their luxuriant qualities, specifically with regard to bearing fruit. One may cite an analogous description in a passage in the Epic of Gilgamesh, which describes a garden-like setting on the mountain called Mashu. The mountain (or perhaps better, *range*) is clearly understood in supernatural terms being described as one

> whose tops [support] the fabric of heaven,
> whose base reaches down to the Netherworld.
> There were scorpion-men guarding its gate.[81]

The following columns set up and describe Gilgamesh's journey into the forest of the mountain. Although many of the details of the text are lost, it describes a place on the mountain that appears to be Gilgamesh's intended destination. The text describes a place where

> NA_4.GUG *na-šá-at i-ni-ib-šá*
> *is-ḫu-un-na-tum ul-lu-la-at a-na da-ga-la* DU_{10}-*bat*
> NA_4.ZA.GÌN *na-ši ḫa-as-ḫal-ta*
> *in-ba na-ši-ma a-na a-ma-ri ṣa-a-a-aḫ*

---

[81]Gilgamesh IX 40–42; translation according to George, *The Epic of Gilgamesh*, 71.

A carnelian tree was in fruit,
hung with bunches of grapes, lovely to look on.
A lapis lazuli tree bore foliage,
in full fruit and gorgeous to gaze on.[82]

The description '(fruit) lovely to look on' (*a-na da-ga-la ṭābat*) and 'fruit ... gorgeous to gaze on' (*in-ba ... a-na a-ma-ri ṣa-a-a-aḫ*) recalls that of the fruit-bearing trees in Gen 2:9: *kol-ʿēṣ neḥmād ləmarʾeh wəṭôb ləmaʾăkāl* 'every tree pleasing in appearance and good for food'. Even more striking is the description of the tree of the knowledge of good and evil, which the text in 3:6 relates 'was a delight to the eyes' (*taʾăwâ hûʾ lāʿênayim*).[83]

Trees, then, figured prominently in the religious experience of ancient Israel, as was true of the general ancient Near Eastern milieu. The emphasis upon luxuriant trees in the description of the garden of Eden appears to stem from the notion that associated trees with the divine presence.

Along with the description of Eden's trees, the Genesis account describes its subterranean water source. Subterranean waters and flowing rivers figure prominently in both biblical and extrabiblical literature. Gen 2:6,10–14, describe the water source of the garden. In v. 6 the narrator relates that 'a mist/flood went up from the earth and watered the entire face of the ground'.[84] According to v. 10, *wənāhār yōṣēʾ mēʿēden ləhašqôt ʾet-haggān ûmiššām yippārēd wəhāyâ ləʾarbāʿâ rāʾšîm*: 'a river flowed forth from Eden to water the garden; from there it divides and becomes four headwaters'. The description of Israelite Eden as the source of the rivers echoes other ancient Near Eastern allusions regarding similar places. The land of Dilmun in

---

[82]Gilgamesh IX; text adapted from Parpola, *The Standard Babylonian Epic of Gilgamesh*, 102, lines 175–78; translation according to George, *The Epic of Gilgamesh*, 75, lines 173–76.

[83] Note also the close association of fruit and the precious stones in the text from Gilgamesh just cited. Precious stones are also found in the Eden narrative. It is true that the jewels mentioned in the Genesis account (*habbədōlaḥ* 'bdellium', *ʾeben haššōham* 'onyx' 2:11) are found *outside* the garden. Still, it is explicitly stated that they are found in places that are watered by the streams which flow forth from it.

[84]For discussion of this verse see H. N. Wallace, *The Eden Narrative* (Harvard Semitic Monographs; Atlanta: Scholars Press, 1985), 73–74.

Sumerian tradition is likewise noted for the subterranean waters which nourish it. At one point, the myth describes how the Sumerian water god Enki commands the release of sweet waters over the land of Dilmun and the salubrious effects that followed. This water is said to emanate from a subterranean source.

> From the "mouth whence issues the water of the earth,"
> brought her sweet water from the earth;
> He brings up the water into her large...,
> Makes her city drink from it the waters of abundance,
> Makes Dilmun (drink from it) the waters of ab(undance).[85]

The action transforms what were once bitter wells into sweet and gives rise to a new name for Dilmun: 'the bank quay house of the land'. These waters fertilize the land and immediately result in the production of crops.

The association of water sources and divine dwellings is seen elsewhere in the ancient Near East. Utnapishtim in the Gilgamesh Epic, and his earlier Sumerian counterpart, Ziusudra, were placed in the abode of the gods at the head of the rivers.[86] In the Ugaritic literature, El's dwelling is said to be ***mbk nhrm qrb ʾapq thmtm*** 'at the sources of the (two) rivers in the midst of the streams of the two deep places'. A similar allusion may be found in Job 38:16.[87]

A similar concept is found in the Hittite myth (itself a version of a Levantine/Canaanite myth) known as "Elkunirsa and Ashertu," which relates that the storm god, Baal, sought out Elkunirsa, who could be found at his tent-dwelling located at the headwaters of the Euphrates River.[88]

---

[85]Lines 55–59. Translation of Kramer, *ANET*, 38.

[86]Cf. Gilgamesh XI 202–205 (George, *The Epic of Gilgamesh*, 97); *Deluge*, lines 252–62 (*ANET*, 44). I do not intend to equate this expression of the divine habitation with a "garden." The head of the rivers is nonetheless highly suggestive of a fertile place.

[87]*mbk* is thought to be related to Hebrew **nēbek*, 'spring' as well as to Ugaritic *npk* 'spring'. The two phrases *mbk nhrm* and *ʾapq thmtm* appear to be paralleled in *nibkê yām* and *ḥēqer-təhôm* in Job 38:16, where Yahweh asks Job, 'Have you entered into the *springs of the sea*, or walked in the *depths of the sea*?'. Wallace, *The Eden Narrative*, 76.

[88]For translation and bibliography see H. Hoffner, *Hittite Myths*, (Atlanta: Schol-

Joel 4:16–18 [E 3:16–18] describes Yahweh in the midst of Zion, his holy mountain. Verse 17 gives these words of Yahweh:

וִידַעְתֶּם כִּי אֲנִי יְהוָה אֱלֹהֵיכֶם
שֹׁכֵן בְּצִיּוֹן הַר־קָדְשִׁי....

> And you will know that I am Yahweh your God
> The one who dwells on Zion, my holy mountain....

In 4:18, the text describes in future terms the following:

> ...a fountain (*maᶜyān*) will go forth from the house of Yahweh (*mibbêt yhwh*).[89]

The significance of Zion as the divine habitation is also seen in Psalm 87. In 87:1, Zion is referred to as 'the holy mountain(s)' (*harərê-qōdeš*). In v. 3 it is explicitly addressed as *ᶜîr hāʾĕlōhîm* 'city of God'. The psalm ends with the words 'the singers like the dancers [say] "all of my springs (*maᶜyānay*) are in you"' (v. 7).

In two passages already cited, we have seen that the prophets ascribed importance to the concept of (subterranean) waters in the understanding of sacred groves. In Ezekiel 31, the extraordinary size and health of the tree in the garden of God is attributed to the fact that 'all its roots went down to abundant waters'. The centrality of the water source is likewise apparent in Isaiah's polemic against the sacred groves (1:29–30; cf. Ps 80:11–12).

Several of the passages to which we have alluded make reference to mountains as the divine setting. This, of course, reflects the interest in the 'high hills'. As we shall see, the garden of Eden is equated with the 'holy mountain of God' in Ezekiel 28.

To summarize then, in the ancient Near East, a mountainous, oasis-like garden setting is one of the traditionally understood dwellings of the gods. In this section, I have argued that the garden of Eden in Genesis (and elsewhere in the Hebrew Bible) was similarly understood to be a divine dwelling, a place where the divine could be encountered unmediated. Just as sacred trees and groves signified the divine presence, so did the garden of Eden. The text presents this

---

ars Press, 1990), 69.

[89]The same imagery is presented concerning Jerusalem in Zech 14:8: 'living waters (*mayim-ḥayyîm*) shall flow out from Jerusalem'.

notion in the phrase *wayyišməʿû ʾet-qôl yhwh ʾĕlōhîm mithalleket baggān lərûaḥ hayyôm* 'and they heard the sound of Yahweh God walking about in the midst of the garden in the cool of the day' (3:8). Here God strolls about almost incidentally, as the original hearers would have expected.

Eden: Archetype for the Temple

In the tradition-historical background of the narrative of Genesis 2–3 there is yet another important idea present. It is an idea not explicitly stated, nor of primary concern for the latest tradent, but an idea that nonetheless exists just beneath the surface. As a locus for divine activity, Eden provides the archetype for the temple as the place where divine and human meet. Just as Eden is the divine dwelling where a human may encounter God unmediated, so also is the temple the divine dwelling where it is possible for a human to encounter the divine unmediated. Before discussing the evidence from the Hebrew Bible, we will begin with one example from Mesopotamian literature. In a bilingual Sumerian-Akkadian text are found the motifs of trees, jewels, subterranean waters, rivers, divine activity, and the temple:

> *ina E-ri-du kiš-ka-nu-ú ṣal-mu ir-bi ina aš-ri el-lu ib-ba-ni*
> *zi-mu-šu uq-nu-u eb-bi ša a-na ap-si-i tar-ṣu*
> *ša* ᵈ*é-a tal-lak-ta-šu ina Eridu ḫegalli ma-la-a-ti*
> *šu-bat-su a-šar er-ṣe-tim-ma*
> *ki-iṣ-ṣu-šu ma-a-a-lu ša* ᵈ*Nammu*
> *[i]-na bīti el-lu ša ki-ma kiš-ti ṣil-la-šu tar-ṣu*
> *ana lib-bi-šu man-ma la ir-ru-bu*
> *ina qi-ri-bi-šu* ᵈ*Šamaš* ᵈ*Tammuz*
> *ina bi-rit pi-i na-r[a-ti] ki-lal-la-an*

> In Eridu the black tragacanth tree grew, in a pure place it was created.
> Its appearance is lapis lazuli stretched out on the Apsu
> of Ea — his promenade in Eridu filled with abundance.
> Its dwelling is the place of the underworld;
> its shrine is the bed of Nammu.
> In the holy temple in which like a forest it casts its shadow,
> into which no one has entered,
> in its midst are Shamash and Tammuz,
> *in between the mouths of the two rivers.*[90]

---

[90]*CT* 16, 46:183–98 = Thompson, *Devils and Evil Spirits* I, 200, lines 183–98. For

There are numerous places in the Hebrew Bible where the temple is presented in the imagery of Eden.[91] This comes as no surprise, for, as we have seen, the luxuriant sacred tree or grove was widely seen as a place of divine habitation. The tabernacle or temple expressed the same notion in more anthropomorphic terms — the divine inhabiting a house. These two concepts were often wed in ancient Near Eastern ideas about the divine. The coalescence of garden and temple is affirmed by the psalmist in 36:8–10:

8 מַה־יָּקָר חַסְדְּךָ אֱלֹהִים
וּבְנֵי אָדָם בְּצֵל כְּנָפֶיךָ יֶחֱסָיוּן׃
9 יִרְוְיֻן מִדֶּשֶׁן בֵּיתֶךָ
וְנַחַל עֲדָנֶיךָ תַשְׁקֵם׃
10 כִּי־עִמְּךָ מְקוֹר חַיִּים
בְּאוֹרְךָ נִרְאֶה־אוֹר׃

8 How precious is your steadfast love, O God.
The children of men take refuge in the shadow of your wings.
9 They feast on the *abundance of your house,*
and you give them drink from the *river of your delights* (*ᶜădāneykā).*
10 For with you is the *fountain of life;*
in your light do we see light.

Here Eden imagery abounds. Feasting on the abundance in the house of God recalls the description of the garden as containing ***kol-ᶜēṣ neḥmād ləmarʾeh wəṭôb ləmaʾăkāl*** 'every tree that is delightful in appearance and good for food' (Gen 2:9). The imagery of a refreshing, flowing water course that 'gives drink" (Hiphil of ***šqh***) recalls the river that 'went forth from Eden to water (Hiphil of ***šqh***; literally, 'give drink to') the garden' (***wənāhār yōṣēʾ mēᶜēden ləhašqôt ʾet-haggān***). The phrase 'river of ***ᶜădāneykā***' is more than suggestive of Eden, although the plural form discourages translating ***ᶜădāneykā*** as a proper

---

text, translation, and discussion of the Akkadian and Sumerian versions, see, in addition to Thompson, S. Langdon, "The Legend of Kiškanu," *JRAS* (1928): 843–48 (see especially 846–47); W. F. Albright, "The Mouth of the Rivers," *AJSL* 35 (1918–19): 161–95; G. Widengren, *The King and the Tree of Life in Ancient Near Eastern Religion* (King and Saviour 4; Uppsala: A.-B. Lundequists, 1951), 5, 6; B. Foster, *Before the Muses*, 2.850.

[91]For an interesting treatment see J. D. Levenson, *Sinai and Zion* (San Francisco: HarperCollins, 1985).

noun, 'river of your *Edens*', in English. Still, the words 'Eden' and 'delight' are identical in Hebrew, and it is difficult to imagine that the ancient hearer would not have noticed this — particularly given the context in which it appears. Finally, *məqôr ḥayyîm* 'fountain of life' highlights the life-giving properties of the water, as is also highlighted in the Sumerian myth concerning Dilmun (cf. Gen 2:10), and which was emphasized in Isaiah's polemic against the sacred groves (1:29–30).

Similar language is found in Jer 17:12–13, another example of garden and temple imagery coalescing:

12 כִּסֵּא כָבוֹד מָרוֹם מֵרִאשׁוֹן
מְקוֹם מִקְדָּשֵׁנוּ׃
13 מִקְוֵה יִשְׂרָאֵל יְהוָה
כָּל־עֹזְבֶיךָ יֵבֹשׁוּ
יְסוּרַי בָּאָרֶץ יִכָּתֵבוּ
כִּי עָזְבוּ מְקוֹר מַיִם־חַיִּים אֶת־יְהוָה׃

12 A glorious throne on high *from the beginning*
is the *place of our sanctuary*.
13 O Yahweh, the hope of Israel,
all who forsake thee shall be put to shame;
those who turn away from [thee]
shall be written in the earth,
for they have forsaken the *fountain of living water*, Yahweh.

Once again the fountain of living water (*məqôr mayim ḥayyîm*) is associated with the sanctuary. Here, Yahweh himself is called the fountain, an idea not far from what we have just seen in Ps 36:10, namely that the fountain of life (*məqôr ḥayyîm*) is *with* Yahweh (*ʿimməkā*). Jer 17:12–13 is highly suggestive of the traditions manifested in the Eden narrative in the references to the beginning (*mērīʾšôn*) and to forsaking Yahweh, a possible reflection on the disobedience of the woman and the man with respect to God's command. The reference to being written in the earth, if it symbolizes death, may also be considered as such.[92] It is particularly interesting that this passage is set within the

---

[92] J. Hyatt (*IB* vol. 5, 956) relates *bāʾāreṣ yikkātēbû* more specifically to the 'ground' or 'dust', which he seems to understand as a trope for death, for he juxtaposes it to being "written forever in the 'book of life' (cf. Exod. 32:32; Isa. 4:3; et al.)." Hyatt implies that to be associated with the eternal temple (v. 12) is equivalent to life.

context of Jeremiah's polemic against the sacred groves. In 17:2 he invokes the saying which is common in Deuteronomistic literature in speaking against worship at these unauthorized locations. He states that the children of Judah 'remember their altars and their Asherim *beside every green tree, and on the high hills*'. In a similar fashion to Isaiah's treatment of this topic (Isa 1:29–30), the prophet continues in vv. 5–8 by stating that the one who trusts in humanity and whose heart turns from Yahweh is like a shrub in the desert, which will 'dwell in the parched places of the wilderness, in an uninhabited salt land' (vv. 5–6). The one who trusts in Yahweh, on the other hand, 'is like a tree planted by water' which 'does not fear when heat comes, for its leaves remain green; and is not anxious in the year of drought, for it does not cease to bear fruit' (vv. 7–8; cf. Ps 1:1–3). Also in this vein, and a poignant statement of all of the ideas we have discussed regarding the significance of the garden of Eden, is Ps 52:10, where the psalmist declares, 'I am like a green olive tree in the house of God' (*ʾănî kəzayit raʿănān bəbêt ʾĕlōhîm*). The psalmist's words capture both the notion of the garden as a locus for divine activity and that of the coalescence of garden and temple. Once again, the garden is expressed through the imagery of a luxuriant tree (*zayit raʿănān*).[93]

The temple is the source of life-giving waters in Ezekiel 47. The water flows from below the threshold of the temple (v. 1). Of this river, the prophet writes:

> As I went back, I saw upon the bank of the river very many trees on one side and on the other. And he said to me, "This water flows toward the eastern region and goes down into the Arabah; and when it enters the stagnant waters of the sea, the water will become fresh. And wherever the river goes every living creature which swarms will live, and there will be very many fish; for this water goes there, that the waters of the sea may become fresh; so everything will live where the river goes. Fishermen will stand beside the sea; from En-gedi to En-eglaim it will be a place for the spreading of nets; its fish will be of very many kinds, like the fish of the Great Sea.... And on the banks, on both sides of the river, there will grow all kinds of trees for food (*kol-ʿēṣ-maʾăkāl*). Their leaves will not wither nor their fruit fail, but they will bear fresh fruit every

[93]The choice of 'olive tree' by the psalmist is significant. Not only is the tree fruit-bearing (as in the Genesis account), but the fruit is among the most significant in the economy of ancient Israel.

> month, because the water for them flows from the sanctuary. Their fruit will be for food, and their leaves for healing. (47:7–9, 12)

Here, the image of the subterranean waters flowing forth to make the salt water fresh is suggestive of traditions underlying the Dilmun myth. The waters will be responsible for producing ***kol-ᶜēṣ-maʾăkāl*** 'all kinds of trees for food' (v. 12) recalling the description of Gen 2:9: ***kol-ᶜēṣ neḥmād ləmarʾeh wəṭôb ləmaʾăkāl*** 'every tree pleasing in appearance and good for food'. The garden imagery begins with the temple and is, in an extraordinary fashion, extended throughout the land.

In the book of Jubilees, Eden is regarded as the prototype of the temple, an identification that "provides the basis for the levitical law which requires abstention from the sanctuary and touching holy things during purification."[94] J. Levison correctly observes that the "author imports levitical ideas into his interpretation of Genesis."[95] In 3:12–13, the angel describes how Eve was brought into the garden after the period of her uncleanness from childbirth:

> And when she finished those...days, we brought her into the garden of Eden because it is more holy than any land. And every tree which is planted in it is holy. Therefore the ordinances of these days were ordained for anyone who bears a male or female that she might not touch anything holy and she might not enter the sanctuary until these days are completed for a male or female.[96]

This association of the temple and the garden, however, is not a novel idea. Rather, as we have seen, the association of temple and garden imagery had long been established in the religious consciousness of Israel; the association of the temple with the garden is a logical extension of the conception of the garden as a place of divine habitation.

---

[94] J. R. Levison, ***Portraits of Adam in Early Judaism: From Sirach to 2 Baruch*** (Journal of the Study of the Pseudepigrapha Supplement Series; Sheffield: Sheffield Academic Press, 1988), 93.

[95] Levison, ***Portraits*** , 93.

[96] Translation of O. S. Wintermute in ***The Old Testament Pseudepigrapha,*** ed. J. H. Charlesworth (2 vols.; New York: Doubleday, 1985), 2.59.

The Primal Human in the Garden: Two Traditions

The primal human was included for a specific purpose in the primeval history but nevertheless comes from a larger pool of images common in ancient Near Eastern thought. The text tells us that he was placed there 'to work it and to keep it' (*ləᶜobdāh ûləšomrāh*), but the narrative as a whole seems to suggest that this is more than a simple aetiology for human work; there is a difference between this presentation of work in Gen 2:15 and in the curse of 3:17–19. Gen 2:15 suggests that the human was placed in the garden for the purpose of keeping it, that is, *for the garden's sake*. This image contrasts sharply with that in 3:17 on two counts: in 3:17 the work characterized by toil takes place outside of the garden; and the ground outside of the garden is cursed. In 2:15 the human could live by means of horticulture (or simply by picking wild fruit) in an extraordinarily fertile environment. Outside the garden, however, he could only continue to live by eating the produce of the cursed earth. The aetiology of toil as the lot of humankind is found to a greater degree in the curse than (if at all) in the original context of his creation.

The Gen 2:15 account expresses quite simply that the primal human was created for service in the garden of God. We have already seen that the garden of Eden was not created *for* the human; rather, it reflects commonly understood notions of a place inhabited by the divine. In Genesis 2, the human is placed in Eden to work the garden which God himself planted; hence, in J's account, the human is created, at least in part, to perform agricultural service for God. This situation is set up by 2:5, in which the creation of the human takes place within the context of the absence of most types of vegetation, and is resolved in 2:7–8.

There are two strands of tradition that appear also in Mesopotamian thought which appear to have played a role in the shaping of the Genesis narrative. According to one, humankind was created to do the labor of the gods. The second involves royal ideology, whereby the king is conceived of as a "gardener." We shall now consider how our text relates to these traditions.

Humans as Laborers of the Gods

A recurring theme in the extant traditions regarding the creation of humanity holds that humanity was created to bear the burden for the gods. This idea is found in the opening sections of the Atrahasis myth (OB version), where we are told that before the creation of humanity, the lesser gods carried out the necessary work, imposed on them by the great gods. The story opens with the following words:

> When gods were man
> They did forced labor, they bore drudgery.
> Great indeed was the drudgery of the gods,
> The forced labor was heavy, the misery too much
> The seven(?) great Anunna-gods were burdening
> The Igigi-gods with forced labor.[97]

To ease the burden of these gods, Ea proposed the creation of humanity to perform their work:

> Their forced labor was heavy, [their misery too much]'...
> [Belet-ili, the midwife], is present.
> Let her create, then, a hum[an, a man],
> Let him bear the yoke [ ],
> Let him bear the yoke [ ],
> [Let man assume the drud]gery of god."[98]

The mother goddess Mami was commissioned to carry out the plan:

> "Let the midwife create...,
> "let man assume the drudgery of god."
> They summoned and asked the goddess,
> The midwife of the gods, wise Mami,
> "Will you be the birth goddess, creatress of mankind?
> "Create a human being that he bear the yoke,
> "Let him bear the yoke, the task of Enlil,
> "Let man assume the drudgery of god."[99]

The nature and purpose of the work is presented in lines 336–39. The ultimate purpose is the production of provisions for the gods.

---

[97] *Atr.* I 1–6. Translation of B. Foster, *Before the Muses*, 1.159.

[98] *Atr.* G, lines d, h–l. Translation of Foster, ibid., 164.

[99] *Atr.* I 189–97. Translation of Foster, ibid., 165.

> They took up ... [ ]
> They made n[e]w hoes and shovels,
> They built the big canal banks,
> For food for the peoples, for the sustenance of [the gods].

This notion appears again, later, in the so-called Babylonian Epic of Creation, Enuma Elish. Here the same motif is used, although the general context is somewhat different. Once again, the work referred to seems to concern the provisions of the gods.[100] In this case, the idea is conceived by Marduk who then proposes to Ea:

> I shall compact blood, I shall cause bones to be,
> I shall make stand a human being, let 'Man' be its name.
> I shall create humankind,
> They shall bear the gods' burden that those may rest.[101]

The notion of humankind being created to bear the burden of the gods is also expressed in a bilingual Sumerian and Akkadian text, which presents the work in similar terms, and centers it around the temple.[102]

> The work of the gods shall be their work
> that they might lay down the border trenches forever,
> place the pickaxe and the basket,
> for the temple of the great gods,
> which is suitable for an awe-inspiring high place
> to group field to field,
> to lay down the border trenches forever....
> to make flourish every sort of plant....
> That they make flourish the grainfield of Anunna,
> to increase the abundance in the land,
> to appropriately celebrate the feast of the gods,
> to pour out cold water,
> in the great dwelling of the gods,
> which is suitable for an awe-inspiring high place.[103]

---

[100]Cf. *EnEl* V 139, where the gods remark, 'let them bring us our daily portions'. It is not clear to whom 'them' refers (see B. R. Foster, *Before the Muses*, 1.383 n. 1).

[101]*EnEl* VI 5–8. Translation of Foster, *Before the Muses*, 1.384.

[102]*KAR* 4. For text and translation, see G. Pettinato, *Das altorientalische Menschenbild und die sumerischen und akkadischen Schöpfungsmythen* (Heidelberg: Carl Winter Universitätsverlag, 1971), 74–77.

[103]Lines 27–51, as translated in ibid., p. 75–76.

These texts make clear that the work alluded to was not the pointless rock-breaking work of a chain gang; rather, it had an explicit divine purpose: the work had the temple as a focal point and the feeding of the gods as its end. What is particularly interesting here is that the hard work signified by the pickaxe and pannier and by the digging of borders and dykes has an agricultural orientation and goal, expressed in the growth of 'all sorts of plants', 'increase' in grain, and 'abundance' in the land. From these Mesopotamian texts we may conclude that according to this tradition, humanity was created to work not simply *in place of* the gods, but *on their behalf* — to do work of direct benefit to them. This work involves the divine dwelling, the temple, and is agricultural in nature.[104]

Of what significance are these Mesopotamian images for understanding the Genesis account? We can be somewhat assured that the description in Genesis is drawn from the same pool of images common throughout the ancient Near East. But at the same time it goes without saying that the two accounts are quite different, at the very least in terms of presentation. What is the relationship between the Mesopotamian accounts and that of Genesis? Westermann apparently finds the differences more illuminating than the similarities. In his estimation, the Genesis account is "stripped" of any mythological connection with the world of the gods, and as such represents a "demythologized" account.[105] The clearly stated motivations of the Mesopotamian accounts for the creation of humanity — to bear the burden of the gods — are not so emphatically stated in Genesis. Still, against the notion of an absolute demythologization, we may observe that the mythological connection to the "world of the gods" lurks in the background in the language of the divine council, which is quite

---

[104]Lambert recognizes the cultic orientation without mentioning the temple reference: "the essentials of the story are that the gods had to toil for their daily bread, and in response man was created to serve the gods by providing them with food and drink. On the last point all the Mesopotamian accounts agree: man existed solely to serve the gods and this was expressed practically in that all major deities at least had two meals set up before their statues each day." W. G. Lambert, "A New Look at the Babylonian Background of Genesis," *JTS* 16 (1965): 298.

[105]Westermann, *Genesis*, 222.

clear in the first person plural words of God (e.g., Gen 3:22; cf. 1:26). The mythological connection is also present in the very concept of the garden, which we have seen was commonly understood to represent the domain of the gods. According to Westermann, in the demythologized Genesis account, the reason that humankind is put to work is "because the living space which the creator has assigned to his people demands this work."[106] This is true, but it is not untrue of the Mesopotamian accounts, where it is a question of *who* will do the necessary work. Again, we can assert that in both cases, God and humankind operate in the same space. To conclude, despite the differences there are similarities between the accounts worth noticing. The primal human performs agriculturally oriented service in the place inhabited by the divine. Although we should be careful not to allow the Mesopotamian accounts to be more than suggestive of the unspoken background of the Genesis account, it would be a mistake to see in Gen 2:8, 15 solely a commentary on agriculture and on the need for humans to work for their own subsistence. The essence is that humankind was created for the service of God. This is clear in the Mesopotamian traditions which depict the work as pertaining specifically to gods and the temple, and this is most likely the image suggested in the Genesis narrative.[107]

The King and the Garden in the Ancient Near East

The second stream of tradition, found in Mesopotamia and which appears to be present in the background of the Genesis account, belongs to the sphere of royal ideology. According to it, the king is conceived of as a gardener.

In Mesopotamia, considerable energy was devoted to the creation and maintenance of special garden enclosures.[108] Garden enclo-

---

[106]Ibid.

[107]For further discussion of parallels between the biblical and Mesopotamian traditions, see T. Frymer-Kensky, "Atrahasis Epic and Its Significance for our Understanding of Genesis 1–9," *BA* 40 (1977): 147–55; I. Kikawada and A. Quinn, *Before Abraham Was: The Unity of Genesis 1–11* (Nashville: Abingdon, 1985), 36–53.

[108]For recent literature see D. Stronach, "The Royal Garden at Parsagadae," *AIO* 475–502.

sures associated with royal palaces appear to be an offshoot of an interest in horticulture on a larger scale, an interest in fertile land in general. Assyrian royal inscriptions express the royal interest in horticulture as far back as Tiglath-Pileser I, who showed concern for collecting plants and trees and establishing gardens outside the capital.[109] With the Sargonid kings of Assyria, the practice of attaching garden enclosures to the palace appears to have begun.[110] There are several well-known examples of gardens associated with royal palaces. One may clearly see evidence of royal interest in gardens in ancient Israelite society as well. In recounting the great achievements of a human in life, Qoheleth, presumably in the guise of Solomon, lists among his achievements: 'I made great works; I built houses and planted vineyards for myself; I made myself gardens and parks and planted in them all kinds of fruit trees. I made myself pools from which to water the forest of growing trees' (Eccl 2:4–6). These acts, with several others listed, make Qoheleth greater than all who preceded him in Jerusalem.

In the book of Esther, we read of a garden building used for special feasting occasions. It is called the *bîtan hammelek*, a structure arguably related to the Assyrian *bītānu*.[111] The interest in gardens was not restricted to kings, but included all people of means.[112]

Part of the significance of garden cultivation was the harvesting of special trees and plants. The Erra Epic makes the following reference to a special wood used for fashioning cult statues: 'I changed the place of the *mēsu* tree and of the *elmēšu* (amber): I did not reveal the new place to anybody.... Where is the *mēsu* tree, the flesh of the gods, the ornament of the king of the universe? That pure tree...whose roots reached as deep down as the bottom of the underwor[ld]: ...through the vast sea waters; whose top reached as high as the sky of [Anum]?'.[113]

---

[109]A. L. Oppenheim, "On Royal Gardens in Mesopotamia," *JNES* 24 (1965): 331.

[110]Ibid.

[111]Esth 1:5 and 7:8. For discussion of the *bītānu* in Mesopotamia and biblical reflexes, see A. L. Oppenheim, "On Royal Gardens in Mesopotamia," 328–33. See also D. J. Wiseman, "Mesopotamian Gardens," *AS* 33 (1983): 137–44.

[112]Wiseman, "Mesopotamian Gardens," 144.

[113]Tablet I, lines 148–53. Translation of L. Cagni, *The Poem of Erra* (Sources and Monographs on the Ancient Near East 1,3; Malibu, Calif.: Undena, 1977), 32.

In Mesopotamia, there is considerable evidence for the cultic significance of gardens. The gods are associated with gardens. Inanna, Enlil, Anu, and Adad all have gardens associated with them.[114] As I have already mentioned, in ancient Israel, evidence for cultic significance of gardens is clearly stated in Isa 1:29; 65:3; and 66:17. There may be a royal aspect to this as well, for Manasseh is said to have been buried 'in the garden of his house, in the garden of Uzza' (*bəgan-bêtô bəgan ᶜuzzāʾ*; 2 Kgs 21:18). The garden was a significant aspect of Mesopotamian and Israelite life, and for this reason, we are justified in seeing its appearance in the Genesis narrative as significant. Still, the extent to which the royal aspect plays a role in the Genesis story is less clear. Perhaps the most important evidence for considering royal overtones is the Mesopotamian use of the terms "gardener" and "farmer" (or "landworker") as royal epithets.

The king as gardener might be explained in the light of the god as gardener. I have already mentioned that there is evidence associating various gods with gardens (or portraying them as possessors of gardens). This might be interpreted in a number of ways. In relation to the king, however, one might suggest that it has to do with a sort of mimetic relationship between god and king. A. S. Kapelrud has demonstrated a similar phenomenon with respect to temple building.[115] Perhaps gardening was also conceived of as a task for gods and kings.

We return now to the biblical text. In Gen 2:8–9 we read:

> Yahweh God planted a garden in Eden in the East. There he placed the human which he had formed. Yahweh God caused to sprout from the ground every tree pleasing in appearance and good for food.

The purpose of the human is introduced in v. 15:

> Yahweh God took the human and put him in the garden of Eden to work it and keep it.

As I suggested at the outset of this discussion, the fact that the garden was "planted" by Yahweh is significant. The deity is presented

---

[114]See D. J. Wiseman, "Mesopotamian Gardens," 143–44 and references.

[115]A. S. Kapelrud, "Temple Building, A Task for Gods and Kings," *Or* 32 (1963): 56–62.

in clear anthropomorphic fashion, making a garden in the logical and understandable way, by planting it himself.[116] The planting of the garden by Yahweh presents a picture of God and the first human operating in the same sphere. Yahweh and the first human appear to be almost side by side, working together. In Ps 104:14, which appears to allude to this tradition, Yahweh once again 'causes grass to sprout for the cattle, and plants for human cultivation'. A plausible way to understand *wayyaṣmaḥ* is as recognition of Yahweh's influence over the cosmos, which is evidenced in every recurrence of new growth. Yahweh is responsible for the fertility of the land. Just as Yahweh is responsible for the creation and maintenance of such fertility, so is the king. The J image of gardening is a *royal* image.

M. Hutter has followed W. Fauth in asserting the ideological connection between garden work and the duties of the ruler in the Near East and in ancient Greece.[117] The cultic orientation of gardens established by kings may be seen as far back as Gudea of Lagash.[118] The conception of the king as gardener is seen most clearly in the epithets of Akkadian and Sumerian rulers, ENGAR/*ikkaru* 'farmer, landworker', and NU-KIRI$_6$/*nukarribu* 'gardener'.[119]

Two narrative accounts have been cited as bearing witness to the connection between the concepts of kingship and gardening. The so-called "Sargon Chronicle," written in the Neo-Babylonian period, tells of the times of the earlier and great dynasty of Sargon of Akkad. It relates that Irra-imitti, a king of Isin, 'installed Bel-ibni, the gardener,

---

[116]The J account is generally recognized as more anthropomorphic than that of P. The description differs greatly from that presented by P in 1:11, where vegetation is created by simple fiat. The use of the Hiphil verb *wayyaṣmaḥ* 'he caused to sprout' in 2:9 certainly has the flavor of the supernatural. It never occurs in the Hebrew Bible with a human subject when referring to horticulture or agriculture.

[117]M. Hutter, "Adam als Gärtner und König [Gen 2:8,15]," *BZ* 30 (1986): 258–62.

[118]T. Jacobsen, *Harps that Once... Sumerian Poetry in Translation* (New Haven and London: Yale University Press, 1987), 435; M. Gothein, *A History of Garden Art* (London and Toronto: J. M. Dent, 1928), 30.

[119]Cf. also n a - g a d a/*nāqidu* (shepherd), or s i p a/*rē'û* (shepherd) which de Liagre Böhl has said "are tangled in an ancient mythological world view" and belong to a "sacral profession circle" (*Opera Minora* [Gronigen: Wolters, 1953], 418). These terms likewise demonstrate the use of common vocations to symbolize higher offices.

on his throne as a 'substitute king' and...placed his own royal crown on his (i.e., Bel-ibni's) head'. Upon the death of Irra-mitti, Bel-ibni 'was elevated to (real) kingship'.[120] The text known as the "Legend of Sargon" also connects the king with gardening. Sargon, who was abandoned by his mother, a high priestess, and placed in a reed basket in the river, was discovered by a gardener:

> Aqqi, the gardener, lifted me out as he dipped his e[we]r.
> Aqqi, the gardener, [took me] as his son (and) reared me.
> Aqqi, the gardener, assigned me to gardening for him.
> While I was gardening, Ishtar granted me her love,
> And for four and....years I exercised kingship.[121]

Here, we find a close connection expressed between gardening and kingship, although its precise nature is difficult to discern. We know nothing about Akki, aside from his profession. The text seems to present gardening as a prelude to kingship, in more than just an incidental way. It is gardening that appears to attract the attention of Ishtar, and the attraction seems to play a role in Sargon's ascension to the throne. The motif of Ishtar's attraction to a gardener also appears in tablet VI of the Gilgamesh Epic, according to which one of Ishtar's love interests was Išullanu, the 'gardener' of Anu, her father.[122] There is no mention of or allusion to kingship, but the text presents the position of gardener with strong cultic overtones. Just the fact that he is the gardener of the high god Anu is significant. He is one of Ishtar's two ostensibly human love interests, the other being 'the keeper of the herd'. Both are presented as officials, offering sacrifices from the area of their oversight. In the Sumerian myth of "Enki and Ninhursag" alluded to earlier, one of the significant (yet mysterious) characters is a gardener.

The notion of the king as gardener in Mesopotamia has also been

---

[120]*ANET*, 267.

[121]Lines 9–13, Tigay's translation: *Evolution of the Gilgamesh Epic* (Philadelphia: University of Pennsylvania Press, 1982), 254. A similar story is told of Gilgamesh by Claudius Aelianus (ca. 200 C.E.) in *De Natura Animalium* (see 3:38–41). Gilgamesh as an infant was thrown from the citadel and caught by an eagle, which set him down in a garden. The text says, 'When the keeper of the place saw the pretty baby he fell in love with it and nursed it; and it was called Gilgamos and became king of Babylon'. See Tigay, *Evolution*, 252–53.

[122]*Gilg.* VI 64 (George, *The Epic of Gilgamesh*, 50).

argued by G. Widengren, who has drawn attention to texts which reveal that the Mesopotamian king was viewed as the guardian of the garden of the gods, and in this sense was conceived to be a gardener.[123]

So what is the significance of the primal human in the garden? The decision to place the human in the garden to work and to keep it comes from the same sphere of ideas which produced the royal epithet "gardener" and texts such as the Sargon Chronicle that feature the gardener as a significant figure. We have already discussed the importance of gardens in the ancient Near East and in Israel, and have seen that their significance cannot be restricted to kingship, or explained by it alone. Rather, it appears that the king as gardener reflects broader ideas of the significance of the garden, specifically with respect to its association with divinity.

It is not necessary to seek an exact correspondence between the Mesopotamian and Israelite materials. The similarities should be suggestive of concerns that are more general than the specific expressions, or that transcend the specific references. Thus, the presence of royal overtones does not require the conclusion that the primal human is a king; rather, the primal human relates to God in the same way that a king does. In fact, it is perhaps less problematic to understand the primal human as standing behind the king as an archetype, rather than to see the king as being *the* explanation of the primal human.

Positioning the primal human as gardener reveals his relationship with God and the divine realm. The special relationship between God and an individual is the foundation of offices of intermediation, in which an individual occupies a special place between humanity and the divine. If it is thought of in these terms, we begin to realize that presence in a place of divine habitation is not the privilege of all humans: in the "mythical" world only the primal humans walk in such a realm; in the "real" world of ancient Israelite society only intermediary figures — prophets, priests, kings, and the like — may do the same.

---

[123]G. Widengren, *The King and the Tree of Life*, see especially chapter 1. See also "Early Hebrew Myths and their Interpretation," in *Myth, Ritual, and Kingship* (ed. S. H. Hooke; Oxford: Clarendon, 1958), 149–203, especially 165–69; I. Engnell, *Studies in Divine Kingship in the Ancient Near East* (Uppsala: Almquist & Wiksell, 1943), 25–30.

The imagery of the king as gardener, then, lies in the prehistory and background of the Yahwist's description of the creation of Adam in Genesis. As we have seen already in chapter 1, the use of royal imagery in the description of the creation of humanity is also known in the P description of Gen 1:26–28. In both the Priestly and Yahwist accounts, this aspect of the tradition's past is ignored, overlooked, attenuated, or subordinated to the point to which it is rendered unrecognizable to the outsider or casual observer. Hence, the royal imagery in Genesis (as well as in Psalm 8) reflects one specific application of the primal human conception. An analogous phenomenon appears in Egyptian texts which regularly hail the reigning king as the image of God, but on occasion extend this royal imagery to common humanity as well.[124] As we shall see, however, Ezekiel views the primal human as a royal figure whose royalty does not extend to all humanity.

The Mesopotamian king as gardener has clear cultic overtones, and the cultic involvement of the king in Mesopotamia has been well documented, as is true in Israel. That we find a similar connection in the ideological background of the Israelite primal human should not be surprising. Kingship is not a notion that may be easily separated from other offices of mediation, and a parallel to Adam's role as servant in the garden may be found in the Israelite priest serving in the temple. We shall discuss this further in chapter 3 and in our conclusion.

The best way to capture the essence of what is expressed in Gen 2:15 is perhaps to consider Adam as divine servant. The place of the service is the very place where the divine is found, and the important concept is the individual working in the presence of God. We shall see how *place* continues to be an important concept with respect to the primal human.

---

[124]E. Hornung, *Conceptions of God in Ancient Egypt* (Ithaca: Cornell University Press, 1982), 138–39. Hornung writes: "Human beings are also 'images of god'; rare but unambiguous references show that this is true of all human beings. In the stories of Papyrus Westcar even a criminal condemned to death is one of the 'sacred herd' of god. In the Instruction for Merikare...mankind, this 'herd of god,' is said to be 'his likeness (*snn*) who came forth from his flesh.'"

Wisdom and the Primal Human

The centrality of wisdom in the Genesis account is expressed in the notion of the tree of knowledge of good and evil. The meaning of the tree has been the subject of considerable discussion. As we shall see, the phrase used to describe it and other considerations suggest very strongly that the tree implies comprehensive knowledge, particularly with respect to that which cannot ordinarily be known by humankind.

There has been considerable confusion over the two trees in the garden. Some commentators have considered the tree of life a later addition to the narrative. That it is not the primary concern is clear in the manner by which the tree of knowledge is discussed, as the only tree. K. Budde pointed out that the body of the narrative deals with one tree (e.g., *haʾēṣ ʾăšer bətôk haggān* 'the tree in the midst of the garden', 3:3; *bəyôm ʾăkolkem mimmennû* 'in the day that you eat from it' 3:5; cf. vv. 11, 12).[125] Only in the introduction (2:9) and the conclusion (3:22, 24) is the tree of life mentioned. C. Westermann has followed this observation, concluding that the original narrative has been expanded at the beginning and end with a motif, the tree of life, from an independent narrative. Westermann found support in the divine reflection of 3:22 which "introduces a new point of view which is foreign to the movement of the narrative." The reflection suggests the principal motif of a separate narrative in which "the man is in search of the fruit of the tree of life so as to preserve life," a motif known also in the Gilgamesh Epic. The tree of life then was worked in as a way of saying that "a similar event was linked with the tree of life as with the narrative." Westermann considers the device in which the narrative of the tree of life speaks through the narrative of the tree of knowledge "an ingenious and intelligent resolution." All this, he concluded, was the work of J, and does not reflect a secondary addition.[126]

The Meaning of the 'Tree of the Knowledge of Good and Evil'

From a grammatical standpoint, the phrase *ʿēṣ haddaʿat ṭôb*

---

[125]K. Budde, *Die biblische Urgeschichte* (Giessen: J. Ricker, 1883), 46–86.

[126]Westermann, *Genesis 1–11*, 212–13.

*wārā*ᶜ may be understood as two nouns in the construct relation (*ᶜēṣ haddaᶜat*) placed alongside the phrase *ṭôb wārā*ᶜ. If *daᶜat* is an infinitive construct, *ṭôb wārā*ᶜ may be construed as the object, thus 'tree of knowing good and evil'.[127]

Regardless of how one explains the phrase grammatically, one must give an account of what is meant by the concept "knowing good and evil." H. Wallace has summarized the positions.[128] Some commentators have understood the knowledge of good and evil as referring to various human faculties. These include the ability to make moral choices (S. Fischer, K. Budde). But it is difficult to explain why God would want to keep this from humans. Others have considered the meaning to be acquiring adult maturity (H. Gunkel, S. R. Driver, W. Staerk, U. Cassuto). Another related position understands knowing good and evil to refer to self-determination. Thus Speiser considered it to entail the "full possession of mental and physical powers." R. de Vaux thought knowing good and evil to refer to moral autonomy in making one's own decision what is good and bad, and as such challenges the order established by God.[129] Similarly, W. M. Clark understood knowing good and evil as belonging to the sphere of responsibility and judgment, reasoned from the judicial contexts in which the phrase appears (e.g., 1 Kgs 3:9, 2 Sam 13:22; 14:17). All of these fall short, however, in explaining Genesis 2–3 according to Wallace. They are particularly inadequate in explaining how the couple became like gods.

Many commentators have taken the tree to signify the knowl-

---

[127]This explanation was rejected by von Rad (*Das erste Buch Mose. Genesis* [Göttingen: Vandenhoeck und Ruprecht, 1949]: 77) but accepted by Wallace (*The Eden Narrative*, 115). For the infinitive construct governing a direct object, see GKC §115d. It may also be understood as a construct chain, whereby *daᶜat* and *ṭôb wārā*ᶜ are in the construct relation; however, the article on *daᶜat* insists otherwise. *ṭôb wārā*ᶜ may modify *ᶜēṣ haddaᶜat* adjectivally ('tree of good and evil knowledge'), but one would expect agreement in gender and definiteness. The best understanding still seems to be 'the tree of knowing good and evil'.

[128]*The Eden Narrative*, 116–32.

[129]E. A. Speiser, *Genesis* (AB; Garden City, N.Y.: Doubleday, 1964), 26. R. de Vaux, *Genèse* (Paris: L'École Biblique de Jérusalem, 1951), 45.

edge of sexual relations.[130] I. Engnell thought it referred to the act of procreation.[131] Wallace has correctly pointed out, however, that the aetiology of marriage and sexual union precedes the act of disobedience. What is more, the curse in 3:16 makes reference only to hardship in childbirth and does not imply that there had been a prohibition on procreation. R. Gordis also made the argument that procreation was the central idea, but he understood procreation as an attempt to procure immortality *vicariously* through one's offspring, as opposed to *personally* through consuming the fruit of the tree of life.[132] He saw the same phenomenon in the Gilgamesh Epic, in a passage that features a discussion between the alewife, Siduri, and Gilgamesh.[133] In this passage, Siduri tells Gilgamesh that death had been set aside for humanity from the beginning, and that he must focus his energies on other pursuits, among which are to "Gaze on the child who holds your hand, let your wife enjoy your repeated embrace."[134] Gordis' position has been criticized in part because Siduri does not, in fact, present children and sex as the only option to personal immortality, which she told Gilgamesh was reserved for the gods alone.[135] Gordis saw in the words "good" and "evil" a reference to "natural" and "unnatural" sexual experience. The most important evidence he cites is 1QSa 1:9–11 which speaks of abstaining from sexual relations until reaching the age of twenty when a person "knows good and evil." Wallace, however, has pointed out that this reference includes intellectual maturity as well

---

[130]See for example H. Schmidt, *Die Erzählung von Paradies und Sündenfall* (Tübingen: J. C. B. Mohr, 1931), 26–28; J. Coppens, *La connaissance du bien et du mal et le péché du paradis* (Gembloux: J. Duculot, 1948); B. Reicke, "The Knowledge Hidden in the Tree of Paradise," *JSS* 1 (1956): 193–202; L. Hartman, "Sin in Paradise," *CBQ* 20 (1958): 26–40.

[131]I. Engnell, "'Knowledge' and 'Life,' in the Creation Story," In *Wisdom in Israel and the Ancient Near East*, ed. M. Noth and D. W. Thomas (VTSup 3; Leiden: E. J. Brill, 1969), 116.

[132]R. Gordis, "The Knowledge of Good and Evil in the Old Testament and the Qumran Scrolls," *JBL* 76 (1957): 123–38.

[133]Gilgamesh X iii 1–14 (OB version).

[134]Lines 11–12; translation from George, *The Epic of Gilgamesh*, 124.

[135]See Wallace, *The Eden Narrative*, 120.

and is not based on purely sexual concerns.[136]

Many commentators have interpreted the tree to signify universal knowledge. So, for example, J. Wellhausen, who characterized it as "that which transcends...the limits of our nature; prying out the secret of things, the secret of the world, and overlooking, as it were, God's hand to see how He goes to work in His living activity, so as, perhaps, to learn His secret and imitate Him. For knowledge is to the ancient world also power, and no mere metaphysic."[137] Likewise, P. Humbert thought this to be knowledge in general. The desire to possess it was the source of hubris which accompanies a break in the original state of dependence on and obedience to Yahweh.[138]

It seems best to follow the line of interpretation established in the modern era by Wellhausen that the knowledge pertains to that which is ordinarily outside the grasp of humanity. Support for this understanding may be found in the phrase *ṭôb wārāᶜ*, which may be taken as an example of merismus. Wallace's discussion appropriately outlines the evidence. He points out that the pair does not always convey merismus, and in several passages *ṭôb* and *rāᶜ* imply mutually exclusive extremes. An example of this is Deut 30:15 which pairs "good" with "life" and "evil" with "death." If the Israelites obey Yahweh's commandments, they will receive life and all of its blessings (v. 16); if they disobey, death will follow (vv. 17–18). Good and evil are similarly opposed in Isa 5:20 which proclaims: 'Woe to those who call the evil good and the good evil' (*hôy hāʾōmərîm lārāᶜ ṭôb wəlaṭṭôb rāᶜ*). They are also said to exchange darkness for light and bitter for sweet.[139] Some passages are ambiguous, such as Gen 31:24,29 and 1 Kgs 3:9. Wallace cites as unequivocal examples of merismus Gen 24:50; Jer 42:6; Lam 3:38; Eccl 12:14; and 2 Sam 14:17,20.

2 Samuel 14 presents one of the better examples for elucidating the meaning of the tree of knowledge in the Genesis account. In this

---

[136]Ibid., 121.

[137]J. Wellhausen, ***Prolegomena to the History of Ancient Israel*** (New York: Meridian Books, 1957), 302.

[138]P. Humbert, ***Études sur le récit du paradis et de la chute dans la Genèse*** (Neuchatel: Secrétariat de l'université, 1940), 83–94, 113–14.

[139]See also Ps 34:15 and Num 24:13.

passage a woman recognizes the efficacy of David's wisdom as king:

כִּי כְּמַלְאַךְ הָאֱלֹהִים כֵּן אֲדֹנִי הַמֶּלֶךְ לִשְׁמֹעַ הַטּוֹב וְהָרָע

> For like the messenger of God is my lord the king discerning good and evil (v. 17).

Similarly in v. 20 she says:

וַאדֹנִי חָכָם כְּחָכְמַת מַלְאַךְ הָאֱלֹהִים לָדַעַת אֶת־כָּל־אֲשֶׁר בָּאָרֶץ׃

> My lord is wise like the wisdom of the messenger of God knowing all that is in the earth.

In these two statements it is apparent that the phrase *lišmōaʿ haṭṭôb wəhārāʿ* is semantically congruent with *lādaʿat ʾet-kol-ʾăšer bāʾāreṣ*; the second statement is most reasonably taken as a restatement of the first. In other words to know *haṭṭôb wəhārā* is to know *kol-ʾăšer bāʾāreṣ*. What is more, although it appears with a different phrase, the use of *daʿat* in this context recalls the description of the tree in Genesis. Perhaps most significant is that the final clause of each verse elaborates upon and explains the initial statement, which in each case likens the king to a divine being. In v. 17 the king himself is likened to the *malʾak hāʾĕlōhîm* and in v. 20, his wisdom is likened to that of the *malʾak hāʾĕlōhîm*. It is difficult then to consider 2 Samuel 14 and Genesis 2–3 unrelated, even though there is insufficient evidence to posit a literary connection between the two passages.[140]

Knowing good and evil, thus, is presented as a divine quality both in Genesis 2–3 and in 2 Samuel 14. The fact that it is a distinguishing characteristic of the king in 2 Samuel 14 is not insignificant, and we shall return to this later.

---

[140]The use of *lišmōaʿ* may also be significant in this regard. Viewed alongside Job 15:8, it may be that this verse is better translated 'hearing/privy to the good and the evil' or to all that can be known, as opposed to 'discerning'. Likewise, *lādaʿat* in v. 20 does not suggest the sense of discernment. The sense does not appear to be reasoning ability, but access to knowledge. As we shall see in chapter 4, it is noteworthy that this apparent royal ideology likens the king to the *malʾak hāʾĕlōhîm* (cf. the prophet in the council).

Out of the Garden: Conflict and the Primal Human

The Text: Genesis 3:1–7

[1]וְהַנָּחָשׁ הָיָה עָרוּם מִכֹּל חַיַּת הַשָּׂדֶה אֲשֶׁר עָשָׂה יְהוָה אֱלֹהִים וַיֹּאמֶר[a]
אֶל־הָאִשָּׁה אַף כִּי־אָמַר אֱלֹהִים לֹא תֹאכְלוּ מִכֹּל עֵץ הַגָּן׃ [2]וַתֹּאמֶר
הָאִשָּׁה אֶל־הַנָּחָשׁ מִפְּרִי עֵץ־הַגָּן נֹאכֵל׃ [3]וּמִפְּרִי הָעֵץ אֲשֶׁר בְּתוֹךְ־הַגָּן
אָמַר אֱלֹהִים לֹא תֹאכְלוּ מִמֶּנּוּ וְלֹא תִגְּעוּ בּוֹ פֶּן־תְּמֻתוּן׃ [4]וַיֹּאמֶר הַנָּחָשׁ
אֶל־הָאִשָּׁה לֹא־מוֹת תְּמֻתוּן׃ [5]כִּי יֹדֵעַ אֱלֹהִים כִּי בְּיוֹם אֲכָלְכֶם מִמֶּנּוּ
וְנִפְקְחוּ עֵינֵיכֶם וִהְיִיתֶם כֵּאלֹהִים יֹדְעֵי טוֹב וָרָע׃ [6]וַתֵּרֶא הָאִשָּׁה כִּי
טוֹב הָעֵץ לְמַאֲכָל וְכִי תַאֲוָה־הוּא לָעֵינַיִם וְנֶחְמָד הָעֵץ לְהַשְׂכִּיל
מִפִּרְיוֹ וַתֹּאכַל וַתִּתֵּן גַּם־לְאִישָׁהּ עִמָּהּ וַיֹּאכַל׃ [7]וַתִּפָּקַחְנָה עֵינֵי שְׁנֵיהֶם
וַתִּקַּח וַיֵּ דְעוּ כִּי עֵירֻמִּם הֵם וַיִּתְפְּרוּ עֲלֵה תְאֵנָה וַיַּעֲשׂוּ לָהֶם חֲגֹרֹת׃

[1]The serpent was more crafty than any of the creatures of the field that Yahweh God had made. He said to the woman, "Has God indeed said, 'Do not eat from any tree of the garden?' [2]The woman responded to the serpent: "We may eat of the fruit of the garden's trees. [3]But from the tree in the middle of the garden, God said 'you shall not eat of it, or you will surely die.'" [4]The serpent responded to the woman, "You certainly will not die! [5]Rather God knows that when you eat from it your eyes will be opened, and you will be like God, knowing good and evil." [6]The woman saw that the tree was good for food, and that it was pleasing to the eyes, and that the tree was desirable to make one wise. She took of it's fruit and ate; and she gave also to her husband with her and he ate. [7]Then, the eyes of the two of them were opened and they knew that they were naked; and they sewed fig leaves together and made loincloths for themselves.

Notes:

1 [a] G and Syr reflect the addition of הנחש, most likely an explicating plus.

The conflict between the primal human and God in the Genesis 2–3 account centers around wisdom. The nature of the conflict is stated and clarified several times: the real problem is the acquisition of the divine prerogative of higher knowledge.

It is generally maintained that disobedience is the chief theme of the narrative. The text certainly presents disobedience as a problem. It reappears in the passage quoted above (3:2–3), but it was introduced already in 2:16–17:

[16]וַיְצַו יְהוָה אֱלֹהִים עַל־הָאָדָם לֵאמֹר מִכֹּל עֵץ־הַגָּן אָכֹל תֹּאכֵל׃ [17]וּמֵעֵץ הַדַּעַת טוֹב וָרָע לֹא תֹאכַל מִמֶּנּוּ כִּי בְּיוֹם אֲכָלְךָ מִמֶּנּוּ מוֹת תָּמוּת׃

[16]And Yahweh God *commanded* the human, saying, "from every tree of the garden you shall freely eat; [17]but from the tree of knowing good and evil you shall not eat. For when you eat from it, you will certainly die."

The use of the verb *ṣwh* 'to command' appears to be significant and conscious. The "command" is both to eat and not to eat, and the phrase *lōʾ tōʾkal* recalls the apodictic legal formulae found elsewhere in the Pentateuch, such as in the Decalogue.[141]

I would argue that the disobedience implied by 2:17 is in fact a pretext for the real concern expressed by the author, the acquisition of higher (divine) knowledge by human beings. The emphases of the text suggest that becoming like God was the main issue in the garden narrative. There are four points in the text which suggest that the *result* of eating from the tree was the problem.

First, when the serpent spoke to the woman, he informed her that they would not die, but that their eyes would be opened and that they would become like God (or gods), knowing good and evil (3:5). His statement did not directly address the significance of the prohibition as a command; rather, he addressed the consequences of the act and the motive behind the prohibition. What the serpent seemed to be saying was that eating the fruit would not directly cause death; it was a point which he made manifestly clear when he introduced the true direct result of eating the fruit, and God's motive for keeping humankind from it. This issue of God's motives raised by the serpent seems to arise from the very nature of the situation: the tree is there and available, yet the humans are not allowed to partake of it. It is possible to conclude that God placed it there to tempt the couple or, in more positive terms, to provide them with the opportunity to con-

[141]The words of the serpent, however, only partially reflect this legal sense. In 3:1 the force of 'command' is softened to 'say' when he asks the woman:

אַף כִּי־אָמַר אֱלֹהִים לֹא תֹאכְלוּ מִכֹּל עֵץ הַגָּן

Has God *said*, "you shall not eat from any tree of the garden?"

The woman likewise uses the verb *ʾmr* 'to say' in her response to the serpent. It may be that *ʾmr* is understood in the sense of a declaration (cf. *kōh ʾāmar yhwh*) bearing equal force to *ṣwh*.

form of their own free will to God's command; but it is equally possible that the tree was understood to be a fixture of the divine garden, which posed immediate tension when the humans were given access to the garden.[142]

Second, in 3:7 the narrator tells of the immediate effects that eating the fruit had on the couple. The first effect is that 'the eyes of the two were opened' (*tippāqaḥnâ*), as the serpent had said. The phrase "open the eyes" occurs elsewhere in the Hebrew Bible, but the other cases provide little help in clarifying the meaning here.[143] "Sensory awareness" and "perception" seem to be the ideas most central in the use of the verb. The use in Genesis 3 appears to be much the same, but it is qualified and explained by the narrator: in 3:5 the serpent says, 'and you will be like or God (or gods) knowing good and evil'.[144]

The first thing of which the man and the woman become aware is their nakedness. In 3:7 we are told that upon eating the fruit, 'the eyes of both were opened ***and they knew*** that they were naked'

---

[142]It would be difficult to deny that a principal interest of the text is to provide an aetiology for human mortality. The narrative flow suggests a significant concern with the effect of eating the fruit and the implications thereof. This concern is as important (if not more so) than the notion of disobedience, which seems to be more of a surface concern. The serpent is called 'crafty' (*ʿārûm*), and is alleged to have 'deceived' (*nšʾ*) the woman. Still, the narrator makes clear that the serpent spoke accurately, at least with regard to their "eyes opening" and "becoming like God." Death did not immediately follow consumption (*bĕyôm ʾăkolkā*); and through this, the narrator suggests that the words of the serpent were true.

[143]The verb *pqḥ* 'to open (the eyes)' occurs twenty times in the Hebrew Bible. (In one case [Isa 42:20] it means 'to open the ears', an apparent reference to spiritual perception.) In the Qal, it is used with reference to both human subjects and God. It is sometimes used literally (2 Kgs 4:35; Job 27:19; Prov 20:13). When used of God, it expresses the notion of his observing something (2 Kgs 19:16=Isa 37:17; Dan 9:18; Jer 32:19; Job 14:3). Zech 12:4 seems to convey the notion of compassion. It is used to express the reversal of blindness, a use which arguably in each case is figurative (Isa 42:7; Ps 146:8; cf. Isa 35:5 Niphal; 2 Kgs 6:20 appears to refer to physical blindness, but is unclear). In 2 Kgs 6:17 the servant of the man of God is enabled to see an array of supernatural forces. In Gen 21:19, God opens the eyes of Hagar, enabling her to see a well of water in the wilderness.

[144]This may also be translated: 'You will be like the gods, those who know good and evil'.

(*wayyēdəᶜû kî ᶜêrūmîm hēm*). The use of the verb *ydᶜ* here suggests that the narrative emphasizes the word "know" and its implications. Prior to consuming the fruit, they lacked shame because they lacked knowledge. The realization of nakedness is a profound discovery, because the concept "naked" exists only in opposition to the concept "clothed." Prior to the advent of clothing there could be no such thing as nakedness. Positing nakedness as the first "post-enlightenment" discovery, then, is a clever device on the part of the narrator, which serves to reveal the extent of their heightened cognition. They then proceed to carry out one of the first acts of culture, the making of clothes (which is probably not insignificant with regard to wisdom).[145]

Third, God's question, 'who told you that you were naked?' (*mî higgîd ləkā kî ᶜērōm ʾāttā*) is particularly interesting. The verb *ngd* means essentially 'to be conspicuous'.[146] In the Hebrew Bible it appears only in the causative stems and hence refers to the disclosure of something previously not known to the addressee. It is frequently used with reference to information known to God alone, and often expresses the revelation of his word. The question 'Who?' appears to be rhetorical, and the narrative does not strongly suggest that God expected an answer. Still, it suggests that they could not have "discovered" by themselves that which they came to know; the issue is not primarily the content of their knowledge, but the nature of their knowledge.

Finally, the veracity of the serpent's statement is confirmed by the narrator in 3:22 when God says, 'the man has become like one of us, knowing good and evil' (*hēn hāʾādām hāyâ kəʾaḥad mimmennû lādaᶜat ṭôb wārāᶜ*). It is this realization that prompts God to take action against the man.

With the eating of the fruit of the tree of the knowledge of good and evil, the primal couple gained access to secrets of the cosmos. They now had knowledge of everything and as a result, had to be stopped. Their fate was sealed, and their expulsion from the garden meant certain death. Only in the garden could one live forever, for only there

---

[145]Cf. R. Oden, *The Bible Without Theology* (San Francisco: Harper & Row, 1987), 94–104.

[146]So BDB, 616.

(presumably) was the tree of life. The necessary measure was to allow the first couple to die.

A similar idea is expressed in the account of the tower of Babel. Genesis 11 relates humanity's attempt to build a city, and a tower 'with its top in the heavens' — likely a metaphorical reference to the pursuit of higher knowledge. The combination of the city, tower, the unity of humanity, and their single tongue points to the issue of human progress, or civilization. The achievements are seen as only the beginning of their aspirations (*haḥillām laᶜăśôt*, v. 6). According to the narrator, Yahweh viewed this potentially unbridled progress as posing a threat to order: "nothing they think to do will be impossible for them' (*lōʾ-yibbāṣēr mēhem kōl ʾăšer yāzəmû laᶜăśôt*, v. 6). The implication is that humanity could learn enough to do anything imaginable, to become, in essence, God, for the secrets of the universe do not belong to humanity. There is a clear problem with humans gaining the divine prerogative of such higher knowledge. In this, there seems to be a tacit recognition that higher knowledge, or wisdom, in the hands of humans is a potentially dangerous thing.

Excursus: Adam and Adapa, Human Wisdom, and Transgression

The Myth of Adapa

The myth of Adapa is known from several fragmentary sources, discovered beginning in 1894. Interestingly enough, the longest and earliest of the fragments comes from fourteenth-century Amarna in Egypt, far from the original Mesopotamian provenance of the story. The other three come from the seventh-century library of Ashurbanipal at Nineveh. The myth opens with a laudatory introduction of Adapa, drawing attention both to his extraordinary wisdom and to his mortality.[147] In the opening scene, Adapa, a priest of Eridu and thus a devotee of Ea, is in a boat on the open water,

[147]See fragment A, lines 1–4 in B. Foster, *Before the Muses*, 1.430–33. Some scholars have interpreted the attribution of wisdom to Adapa's patron god, Ea, but this seems unwarranted given the essence of the story, and the nature of other literary references to Adapa.

fishing, when the South Wind threatens disaster. Adapa curses the wind, an action which proves effectual, for its 'wing' is 'broken'. This action greatly disturbs the high god, Anu, who summons Adapa before his presence in heaven.[148] Before Adapa goes before Anu, however, he is given some curious words of advice by Ea. The advice includes dressing in mourning clothes to curry the favor of the attendants of Anu.[149] The strangest part of the advice then follows: Adapa is told to accept anointing oil and a new garment when offered, but to refuse the offer of water and food, which he calls the *mê mūti* 'water of death' and the *akalu ša mūti* 'food of death'.[150] This is indeed strange, for when he arrives before Anu, he is offered and accepts the oil and new garment; but rather than being offered the *mê mūti* and the *akalu ša mūti*, he is instead offered the *akal balāṭi* 'food of life' and the *mê balāṭi* 'water of life', which he refuses. By doing so, he forfeits his opportunity for immortality. Adapa is then returned to earth.[151]

At first, the Adapa myth seems to have little in common with the myth of Adam in Genesis 2–3, and for this reason, many scholars have refused to recognize any substantial connection between the two. There are, however, a number of significant points of contact that have been recognized and which are important to our investigation.

The most basic of these, perhaps, is the similarity in the names. Gen 2:7 describes the creation of humanity in these words: *wayyîṣer yhwh ʾĕlōhîm ʾet-hāʾādām ʿāpār min-hāʾădāmâ* 'Yahweh God formed the human of dust from the earth'. The name for human comes from *ʾădāmâ*, 'ground', and it is the difference of the bilabials *m* / *p* which separates this from Adapa. Whether Adam derives etymologically from "ground" or "red" is less important than the fact that it means 'human'. The etymology, moreover, is less significant than the similarity, for the generating of folk etymologies was a common practice in the ancient Near East. Moreover, E. Ebeling reported the occur-

---

[148]Fragment A, lines 15–22; B, lines 1–15 (ibid.).

[149]Fragment B, lines 16–41 (ibid.).

[150]Ibid., lines 36–37.

[151]Ibid., lines 76–85.

rence of *a-da-ap* with the definition 'man'.[152] Unfortunately, his reference ("*a-da-ap* ist nach unveröffentl. Vokabular = Mensch") does not give any other information, most importantly, the word used there for 'man'.[153] Unlike *ʾādām*, however, there is no evidence that the association between the name Adapa and humanity was commonly made. Adapa appears to have retained his individuality, whereas Adam, at least in the earlier Israelite material, appears to be less significant as an individual. As we shall see in the chapters to follow, however, this is not quite the case.

Some scholars have seen a similarity between the two characters in that both Adam and Adapa were "tested" with food.[154] The role of the fruit of the tree of knowing good and evil in Genesis is admittedly somewhat different from that of the bread and water of life in the Adapa myth, and as I have stated above, there is little evidence to support the notion that the prohibition against eating from the tree of knowing good and evil in Genesis was a "test." Nonetheless, it is not clear that the offer of the food and drink of life in the Adapa myth constitutes a test either.

An interesting part of the Adapa myth comes in the fragment from Amarna. Upon discovering Adapa's ability to break the wing of the south wind, Anu asks, 'Why did Ea disclose what pertains to heaven and earth to an uncouth mortal?'.[155] This concern over the revelation of divine information to a human recalls Ea's disclosure of divine knowledge to the mortal, Utnapishtim, called Atrahasis, in the flood tablet of the Gilgamesh Epic, and the subsequent decision with

---

[152]Ebeling, *Tod and Leben: nach den Vorstellungen der Babylonier* (Berlin and Leipzig : Walter de Gruyter, 1931), 27, note a.

[153]Equally unfortunate is that many scholars currently invoke Ebeling's comment, largely because it appeared in *ANET* (101).

[154]See, for example W. H. Shea, "Adam in Ancient Mesopotamian Traditions," *AUSS* 15 (1977): 27–41.

[155]*ammīni ea amīlūta lā banīta ša šamê u erṣeti ukillinši.* Fragment B, lines 57–58 (see Heidel, *The Babylonian Genesis,* [Chicago: University of Chicago Press, 1963], 151). Text from S. Izre'el, "Some Thoughts on the Amarna Version of Adapa," in *Mésopotamie et Elam. Actes de la XXXVIème Rencontre Assyriologique Internationale* (Mesopotamian History and Environment Occasional Publications 1, Ghent: University of Ghent Press, 1989), 211–20, especially 218.

which the gods were faced — what to do with a human who had attained divine knowledge (cf. Gen 3:11, 22–24). According to that account, the gods had previously decreed that humanity should be wiped out from the face of the earth by means of a flood. Ea managed to communicate this impending doom to Utnapishtim and warned him to escape by building a boat. After the furious deluge, so severe that it frightened the gods themselves, the storm-god Enlil discovered the boat, and that a human had avoided catastrophe by knowing the plans of the gods. The enraged god exclaims, '[From] where escaped this living being? No man was meant to survive the destruction!'[156] To this, Ea responds:

> It was not I disclosed the great gods' secret:
> Atrahasis I let see a vision,
> and thus he learned the secret of the gods.
> And now, decide what to do with him![157]

The fact that divine information had fallen into the hands of a human created a problem for the gods, forcing them to decide what to do. The decision was to grant Utnapishtim immortality and to remove him to the place of divine habitation:

> In the past Utnapishtim was a mortal man,
> but now he and his wife shall become like us gods!
> Utnapishtim shall dwell far away, where the rivers flow forth![158]

In the Sumerian analog, Ziusudra, the sole survivor of the flood, is granted eternal life, following the same circumstances:

> Life like (that of) a god they give him,
> Breath eternal like (that of) a god they bring down for him.
> Then, Ziusudra the king,
> The preserver of the name of vegetation (and) of the seed of mankind,
> In the land of crossing, the land of Dilmun, the place
> where the sun rises, they caused to dwell.[159]

---

[156]Gilgamesh XI 174–75; George, *The Epic of Gilgamesh*, 95.

[157]Gilgamesh XI 195–97; translation adapted from George, *The Epic of Gilgamesh*, 95.

[158]Gilgamesh XI 202–205; George, *The Epic of Gilgamesh*, 95.

[159]*Deluge*, 252–59; *ANET*, 44.

In the Adapa myth, a similar situation is presented. Having realized Adapa's powers, which, no doubt, proceeded from his wisdom described in the beginning of the myth, Anu declares:

> What, for our part, shall we do for him?
> Bring him food of life, let him eat.[160]

Here, the response to Adapa's procuring of divine understanding is the same: an offer of eternal life. Adapa, however as we have already seen, had been instructed by Ea not to partake of the food or water. Accordingly, he refused, much to Anu's apparent amazement. The text, after expressing the god's bewilderment, relates his subsequent decision: 'let them take him and [ret]urn him to his earth'.[161] In the light of the flood narratives discussed above, the returning of Adapa to earth is unexpected; it is a twist on the motif of granting life and placement in a divine locale.

Like the Mesopotamian flood heroes and like Adapa, when the first couple procure divine knowledge, the gods must then decide what to do with them. But unlike the flood heroes, what Adam and Adapa share here is that they are kept from eternal life and evicted from the divine locale: Adam is evicted from the garden and Adapa is returned to earth.

It is perhaps safest to conclude that the myth of Adapa has very little to do with that of Adam in a direct sense. There is little if any evidence of direct contact between the myths. But it is appropriate to pause here briefly to consider the issue of borrowing and common stock. Ever since the discovery of Mesopotamian literary traditions, there has been intense interest in parallels with the biblical traditions. Indeed the interest in biblical parallels fueled the study of Assyriology in its early days, and the writings of the pan-Babylonian school represent one of the more infamous witnesses to and outgrowths from this fact.[162] Although the trend recognized today as "parallelomania" has been tempered somewhat by recent discussions of method, there is

---

[160]Foster, *Before the Muses*, 433.

[161]Line 85, ibid., 433.

[162]Note particularly the controversy generated by Franz Delitzsch's *Bibel und Babel* (Leipzig: J. C. Hinrichs, 1902) and other lectures.

still something to be said concerning the similarities and differences in the literature of the ancient Near East, for it comes from an environment in which regions were linked together by trade, and shared a common linguistic heritage and the phenomenon of cuneiform culture. W. G. Lambert has argued that the period in which the transmission of ideas between the east and west (particularly the movement west) took place was the Amarna period, when the Babylonian language and cuneiform script was the normal means of international communication between Egypt and the Persian Gulf.[163] Fragments of Mesopotamian myths have been found in the Hittite capital of Boghazköy (including the Gilgamesh Epic). A fragment of the Atrahasis Epic has been found at Ras Shamra (ancient Ugarit). Even in Palestine, a fragment of the Gilgamesh Epic has been found, at Megiddo. There are several examples of Mesopotamian texts at Amarna, and as we have seen, one of those is a part of the Adapa Myth. Among the many scholars who have addressed this, Lambert supports the theory that there was considerable cultural borrowing taking place among the various cultures and that oral tradition was responsible for a considerable amount of dissemination.[164] It is most likely the case, then, that formative Israel possessed and developed a stock of oral tradition that was a part of the environment into which it gradually came into existence as a people and as a nation. Hence, it is plausible to assume some knowledge of this Adapa myth or of an allomorph of it in ancient Israel. Given the passage of time between the Late Bronze period and the Iron IIa period, when the Israelite traditions began to assume written form, one would be surprised to find a close parallel of an entire myth, but a comparison of the Israelite Adam and the Mesopotamian Adapa reveals sufficient commonalities to suggest that we are dealing with the same *type* of figure.

Adapa is a type of primal human figure in the way in which he stands between the gods and humanity. Adapa is not a god. He is some manner of semi-divine, enjoying a special relationship with the

---

[163]W. G. Lambert, "A New Look at the Babylonian Background of Genesis," *JTS* 16 (1965): 287–300.

[164]Lambert, "A New Look," especially 299–300 and references.

gods. He is to be associated with humanity by the reference from the lexical text mentioned above (*a-da-ap* = man), and by the emphasis placed on his mortality in the myth. In the myth he is called the 'seed of man', which doesn't necessarily set him apart explicitly as the first human, but it unequivocally reveals his humanity.[165]

Adapa and the *apkallu* Tradition

In Mesopotamian traditions, Adapa is a sage and one who brings higher knowledge to humanity. In a list of sages and kings from the city of Uruk, Adapa (called U'an) appears as the first of seven antediluvian *apkallu* scribes, each associated with a king.[166] The list continues with more kings associated with *ummānu* scribes, providing evidence of the close relationship between the terms *apkallu* and *ummānu*.[167] This list compares with an another list, mentioned above in chapter 1, that makes reference to 'seven brilliant apkallu's, *purādu*-fish of the sea', and begins, '[Adapa,] the purification priest of Eridu [...] who ascended to heaven'.[168] He appears as a culture bearer in a later text drawn from this tradition, in which the Babylonian priest Berossus writes:

> In that first year a beast named Oannes appeared from the Erythraean Sea in a place adjacent to Babylonia. Its entire body was that of a fish, but a human head had grown beneath the head of the fish and human feet likewise had grown from the fish's tail. It also had a human voice.[169]

According to Berossus, Oannes (most likely the U'an of the Uruk List)

---

[165]Fragment D line 12. For a discussion of this phrase, see A. Heidel, *The Babylonian Genesis*, 152 note 33.

[166]For this text, commonly known as the "Uruk List," see J. van Dijk, "Die Inschriftenfunde," *UVB* 18 (1962): 45.

[167]On the interchangeability of the two terms, see B. Foster, "Wisdom and the Gods in Ancient Mesopotamia," *Or* 43 (1974): 345 n. 4.

[168]Text *LKA* 76 published by E. Reiner, in "The Etiological Myth of the Seven Sages," *Or* 30 (1961): 1–11.

[169]I 1.5. Translation of S. Burstein, *Berossus the Chaldean* (Sources from the Ancient Near East 1,5; Malibu, Calif.: Undena Publication, 1978), 13. See also that of B. Foster, "Wisdom and the Gods in Ancient Mesopotamia," *Or* 43 (1974): 347.

> gave to men the knowledge of letters and sciences and crafts of all types. It also taught them how to found cities, establish temples, introduce laws and measure land. It also revealed to them seeds and the gathering of fruits, and in general it gave men everything which is connected with civilized life.[170]

This type of higher learning, which Oannes/Adapa brings to humanity, is reminiscent of the concerns expressed in the tower of Babel account discussed above; and in the J account in Genesis 2–3, the primal couple's partaking of the tree of knowing good and evil may be an allomorph of the notion of the introduction of higher knowledge to humankind. Adapa is most commonly given the epithet *apkallu*, although he is also called *ummānu*.[171] His role as culture bearer in the form of Oannes reveals a special interest in humanity and reflects the early figures of the genealogies. His form as one of the seven sages, a composite fish-like animal, reveals his nature as one who stands between the gods and humanity.

Significant for this study is the veneration of Adapa, both within and outside the myth, for his wisdom. He is frequently invoked in contexts appealing to wisdom. The myth opens by expressing the extent of his wisdom:

> Un[derstanding]
> His utterance can command, like the utterance [of Anu],
> He made him perfect in wisdom, revealing (to him) the designs of the land.
> To him he granted wisdom, eternal life he did not grant him.
> In those days, in those years, the sage, the citizen of Eridu,
> Ea created him as ... among men,
> The sage (*apkallu*) whose pronouncement no one gainsaid,
> Able one, his perception of what pertains to the Anunna-gods was vast.[172]

---

[170] I 1.5. Burstein, *Berossus the Chaldean*, 13–14. On the identification of Oannes with Adapa see B. Foster, "Wisdom and the Gods in Ancient Mesopotamia," 346–47; W. G. Lambert, "A Catalogue of Texts and Authors," *JCS* 16 (1962): 59–77, especially 73–74.

[171] For references, see Reiner, "The Etiological Myth of the Seven Sages," 8. This priestly orientation of the primal human figure is present in the account of the primal human in Ezekiel 28 (see chapter 3 below).

[172] Fragment A, lines 1–8; Foster's translation, *Before the Muses*, 430. The fragment is from the library of Ashurbanipal (mid-seventh century).

What is relevant for our purposes is the emphasis upon the extent of his wisdom. He is a human character, yet his knowledge is equal to that of the gods.

In a seventh-century letter to Ashurbanipal, the writer relates, 'In a dream the god Ashur said to (Sennacherib) the grandfather of the king, my lord, "O sage! You, the king, lord of kings, are the offspring of the sage and of Adapa.... You surpass in knowledge Apsu and all craftsmen"'.[173]

This wisdom of Adapa is variously manifested. He is called an *apkallu*, a 'sage'. As such, he is an important figure in the wisdom tradition of Mesopotamia. He is the first of the antediluvian *apkallu* sages, and as such is identified with the composite human-fish culture bearer, Oannes, by Berossus.[174] According to Reiner, he is a vizier-type figure, appearing in a list of similar figures associated with kings.[175] The type of knowledge associated with Adapa is esoteric and linked to the cultic aspects of magic.[176] Adapa's higher knowledge is expressed in the "Verse Account of Nabonidus," where he represents the major standard of knowledge: 'not (even) learned Adapa knows...'.[177] As a sage, he is considered responsible for writing an astrological omen series.[178]

Also significant for this study is the fact that he is a priest in the myth; in another text, he is identified as a purification priest (*iššipu*).[179] The presentation of Adapa as both sage and priest draws attention to him as a mediator figure. He is not simply "a priest" or "a sage" from the historical past; rather, he is the quintessential wisdom figure, intimately associated with the divine, yet inextricably woven with humanity. The priestly orientation of Adapa and traditions linking him with

---

[173]*ABL* IX 923; *ANET*, 450.

[174]See S. Burstein, *Berossus the Chaldean*, 13.

[175]E. Reiner, "The Etiological Myth of the Seven Sages," especially 7–10.

[176]P. Michalowski, "Adapa and the Ritual Process," *RO* 41 (1980): 77–82.

[177]*ANET*, 131.

[178]This series is also mentioned in the Verse Account of Nabonidus. *ANET*, 314. Some have asserted that the basic domain of Adapa is magic (Michalowski, "Adapa and the Ritual Process," 78).

[179]*LKA* 76, see E. Reiner, "The Etiological Myth of the Seven Sages."

the sea and the "pure *purādu*-fish" appear to have been manifested in the cult. Evidence for this may be seen in an object found at Assur, dated to the period of Sennacherib. The object is a large rectangular tank, most likely a ritual basin. It is decorated with pictures of a god holding a vase flowing with four streams of water (most often identified with Ea); next to each god is a man, undoubtedly a priest, wearing a fish-skin holding a situla.[180]

In terms of the overall significance of the myth of Adapa, P. Michalowski has argued that it represents the institutionalization of the magic arts. The issue is the recognition of the extent of Adapa's power and the fact that limits must be set to it, the result of which is the creation of the institution of magic. Key to this interpretation is that in becoming institutionalized, this power became restricted to a small group of individuals, those involved in the practice of *āšipūtu*.[181] According to Michalowski,

> the story signifies a transition into the crafts of the god Ea, a marginal world inhabited by magicians, demons and, perhaps, sages. All of these beings are marked as being somehow set apart from the regular classificatory scheme of things. The demon...and the magician...play a dialectical role with one foot in the sphere of the sacred and one in the profane world. Both of them have to cross the threshold between these two parts of the universe.... Thus only Adapa and the demon Pazuzu break the wings of winds, a narrative device which may perhaps serve to signal the connections between these types of marginal figures.[182]

It is this marginality that I find significant in Adapa, and that I believe he shares with the Israelite primal humans. Michalowski explains this institutionalization as occurring in the myth, which he describes as an expression of the well-known ritual process in its states of *separation*, *margin*, and *reaggregation*. Although it is difficult to find evidence in ancient Israel of this institutionalization conceptualized in the same way, a similar functional role may be found with respect to the Israelite primal human. We shall discuss this further in the conclusion.

---

[180]See A. Parrot, *The Arts of Assyria* (New York: Golden, 1961), 74.

[181]*āšipūtu*, generally speaking, refers to forms of what is normally called "exorcism." See *CAD* A/2, 435.

[182]Michalowski, "Adapa and the Ritual Process," 81.

Part II

Indirect Attestations

The Primal Human as Analogy

3. The King of Tyre as the Primal Human
(Ezekiel 28:11–19)

IN EZEK 28:11–19 WE ENCOUNTER a presentation of the primal human that is fundamentally different from what we observed in the traditions of Genesis. It is a prophetic oracle built on allusions to the primal human. In it, mythological elements are invoked to make a point. The primal human becomes an expression of another protagonist — the king of Tyre, the putative focal point of the oracle. The text tells us something of the myth of the primal human in an indirect way. Here the primal human is not cast within simple linear history, or as the past progenitor of humanity responsible for the ills of the race. Rather, he emerges in the mind of the tradent as a mythical paradigm whose identity coalesces with that of the Tyrian monarch. The king of Tyre is not simply "like" the primal human, he *is* the primal human.

The Text: Ezekiel 28:11–19

[11]וַיְהִי דְבַר־יְהוָה אֵלַי לֵאמֹר׃ [12]בֶּן־אָדָם שָׂא קִינָה עַל־מֶלֶךְ צוֹר
וְאָמַרְתָּ לּוֹ כֹּה אָמַר אֲדֹנָי יְהוִה אַתָּה חוֹתֵם תָּכְנִית מָלֵא חָכְמָה וּכְלִיל
יֹפִי׃ [13]בְּעֵדֶן גַּן־אֱלֹהִים הָיִיתָ כָּל־אֶבֶן יְקָרָה מְסֻכָתֶךָ אֹדֶם פִּטְדָה
וְיָהֲלֹם תַּרְשִׁישׁ שֹׁהַם וְיָשְׁפֵה סַפִּיר נֹפֶךְ וּבָרְקַת וְזָהָב מְלֶאכֶת תֻּפֶּיךָ
וּנְקָבֶיךָ בָּךְ בְּיוֹם הִבָּרַאֲךָ כּוֹנָנוּ׃ [14]אַתְּ־כְּרוּב מִמְשַׁח הַסּוֹכֵךְ [ו]נְתַתִּיךָ
בְּהַר קֹדֶשׁ אֱלֹהִים הָיִיתָ בְּתוֹךְ אַבְנֵי־אֵשׁ הִתְהַלָּכְתָּ׃ [15]תָּמִים אַתָּה
בִּדְרָכֶיךָ מִיּוֹם הִבָּרְאָךְ עַד־נִמְצָא עַוְלָתָה בָּךְ׃ [16]בְּרֹב רְכֻלָּתְךָ מָלוּ
תוֹכְךָ חָמָס וַתֶּחֱטָא וָאֲחַלֶּלְךָ מֵהַר אֱלֹהִים וָאַבֶּדְךָ כְּרוּב הַסֹּכֵךְ מִתּוֹךְ
אַבְנֵי־אֵשׁ׃ [17]גָּבַהּ לִבְּךָ בְּיָפְיֶךָ שִׁחַתָּ חָכְמָתְךָ עַל־יִפְעָתֶךָ עַל־אֶרֶץ
הִשְׁלַכְתִּיךָ לִפְנֵי מְלָכִים נְתַתִּיךָ לְרַאֲוָה בָךְ׃ [18]מֵרֹב עֲוֺנֶיךָ בְּעֶוֶל
רְכֻלָּתְךָ חִלַּלְתָּ מִקְדָּשֶׁיךָ וָאוֹצִא־אֵשׁ מִתּוֹכְךָ הִיא אֲכָלַתְךָ וָאֶתֶּנְךָ לְאֵפֶר
עַל־הָאָרֶץ לְעֵינֵי כָּל־רֹאֶיךָ׃ [19]כָּל־יוֹדְעֶיךָ בָּעַמִּים שָׁמְמוּ עָלֶיךָ בַּלָּהוֹת
הָיִיתָ וְאֵינְךָ עַד־עוֹלָם׃

> [11]And the word of Yahweh came to me: [12]"Son of Man, lift up a lamentation over the king of Tyre and say to him, Thus says Yahweh God: 'You were a seal of resemblance, full of wisdom and perfect in beauty. [13]You were in Eden, the garden of God; every precious stone was your covering: sardius, topaz, and moonstone; gold topaz, carnelian, and jasper; lapis lazuli, garnet, and emerald; and of gold was (its) workmanship.[a] Your timbrels and your pipes[b] were established on the day you were created. [14]I placed you with the anointed guardian cherub; you were on the holy mountain of God. Among stones of fire you walked around. [15]You were blameless in your ways from the day you were created until iniquity was found[c] in you. [16]In the abundance of your trade you filled yourself[d] with violence and you sinned. I cast you as a profane thing from the mountain of God and the guardian cherub drove you[e], from the midst of the stones of fire. [17]Your heart was proud because of your beauty; you corrupted your wisdom for the sake of your splendor. I cast you upon the earth; I set you before kings so they could look upon you.[f] 18By the multitude of your iniquities in the unrighteousness of your trade you profaned your sanctuaries. So I sent fire from the midst of you. It consumed you and I made you dust on the earth in the eyes of all who see you. [19]All who know you among the peoples were awestruck over you. You became an object of terror[g] and you ceased to exist forever.[183]

Notes:

[a]Reading מלאכה for MT singular construct form מלאכת.

[b]Taking MT בך 'with you' as a dittograph arising from the sequence נקביך ביום.

[c]The disagreement between the verb *nimṣāʾ*, which is masculine and the subject *ʿawlātâ*, which is feminine is admissible on the grounds that it frequently occurs when predicate precedes the subject (GKC §145o).

[d]Lit. 'your midst' The word *mālû* is from the root *mlʾ*. For the omission of the final *ʾaleph* see GKC §75qq. See also R. Meyer, "Zur Sprache von 'Ain Feschcha," *TLZ* 75 [1950]: 721f against Bauer-Leander §54f, g. The best reading מל(א)ת 'you filled' is suggested by Greek επλησας.

[e]This may alternatively read 'I drove you O guardian cherub', a reading which I do not adopt here, but is highly suggestive concerning variant early interpretations of this passage.

[f]Lit. 'to look at you'.

---

[183]This oracle is given as a *qînâ*, and presumably was originally conceived with some measure of metrical regularity. Nevertheless, the MT is so difficult with respect to text, grammar, and semantics that the editors of *BHS* opted to forgo metrical arrangement. For convenience of discussion, I have elected to do the same. The latter portions of the oracle yield more obvious signs of parallelism, and I will comment accordingly.

[g]Following Zimmerli (2.88) for the difficult *ballāhôt hāyîtā* 'you were calamities.'

Several observations suggest that the imagery invoked by the prophet is, in fact, primal human imagery.[184] This second oracle in Ezekiel 28 (vv. 11–19) has often been cited in discussions of the first human and the paradise narrative, despite the absence of an explicit and indubitable reference to a first human. The humanity of the protagonist is expressed, although not *explicitly* stated.[185] It is expressed in the reduction of the protagonist 'to ashes' (v. 18). The protagonist lives in the divine habitation (*bəᶜēden gan ʾĕlōhîm hāyîtā;* v. 13), and the oracle makes allusion to the apparent divine knowledge of the protagonist, in the reference to the stones of the high priestly breastpiece (v. 13), which suggests that here too, wisdom is understood within the context of the protagonist's humanity; the stones signify divine knowledge in the hands of a human.

Several considerations may account for why the primal human is invoked to elucidate the situation of Tyre. Tyre possesses an almost uncanny ability to create wealth. This ability is appropriately construed as "wisdom," and in the second oracle wisdom is embodied in the stones of the high priestly breastpiece. The beauty of the king is presented as that of the primal human. Tyre's wealth and "physical" beauty are also analogous to Eden, the archetype of the temple. The offenses of the primal human and of the king of Tyre involve wisdom. The image is suggestive of the expelled primal human as an excommunicated priest.

Structure and Analysis

From the standpoint of prophetic form, the passage contains a historical retrospect (11–15) and a pronouncement of judgment (16–19).

---

[184]A. Bentzen (*King and Messiah* [Oxford: Basil Blackwell, 1970], 42) accepted that both Ezek 28:1–10 and Ezek 28:11–19 refer to the "First Man."

[185]In the first oracle (vv. 1–10), the protagonist is explicitly identified as a human (*ʾattā ʾādām;* v. 2). This human protagonist lives in the divine habitation (*môšab ʾĕlôhîm yāšabtî;* v. 2). Allusion is made to the apparent divine knowledge of the protagonist, which seems to be understood within the context of the protagonist's humanity (*tittēn libbəkā kəlēb ʾĕlōhîm;* v. 6).

The entire oracle laments events in the past. From the standpoint of the mythic complex of ideas involved, a reading of vv. 12b–19 yields the following structure:

1. Previous Condition (12b–14)
2. Transition to New Condition (15)
3. New Condition and Consequence (16–18)
   - A. First Statement (16)
   - B. Second Statement (17)
   - C. Third Statement (18)
4. Conclusion (19)

Despite the difficulties in discerning the poetic construction of the passage, it nonetheless yields a quite recognizable general structure that features a contrast of states or conditions. It divides at the approximate midpoint: the first half deals with the figure in a previous condition, the second half discusses the figure with respect to its new condition. Within the second section, each verse again divides in half: the first half states the new condition, the second half states the consequences applied by Yahweh. Verse 15 stands as a transitional verse between the two sections, and describes the movement of the figure from one state to the other. The two states may be summarily described as cleanness and uncleanness. We will examine these two states under the headings "the exalted state of the primal human" and "the 'fallen' state of the primal human."

The Exalted State of the Primal Human (12b–14)

Verses 12–14 describe the glorious exalted state of the primal human. There are some difficulties in the language that present obvious problems, but the state of exaltation is nonetheless clear. The information given in these verses suggests both royal and priestly imagery. The picture I believe we are given is that of the primal human, endowed at his creation with royal and priestly accouterments, a common topos in the ancient Near East.[186] As in the Genesis 2–3

[186]On this topos, see J. Tigay, *Evolution of the Gilgamesh Epic*, 153–54.

account, the primal human is set within the context of the divine habitation and cohabits with the divine.

The seal: authority and kingship by proxy (12b)

MT *ʾattâ ḥôtēm toknît* presents both lexical and grammatical problems, and, as a result, has resisted scholarly consensus in interpretation.[187] Particularly distressing for our purposes is the fact that, because it comes first, the phrase may represent our best clue to the person and character of the mythological figure. Despite the difficulties presented by the text, however, I believe there is sufficient biblical and extrabiblical evidence to warrant the following judgment: the phrase is a reference to the imagery and ideology of the royal seal. After examining some of the difficulties posed by the text, and some linguistic possibilities for translating the phrase, I will discuss the interpretation of the primal human as a seal.

The word *toknît* appears elsewhere only in Ezek 43:10 where, despite the corrupt text, it seems to signify pattern or model.[188] It may be that the original Hebrew text of Ezek 28:12 read *tabnît* 'image' rather than *toknît*, and the MT reading arose through a graphic confusion of *k* and *b*. But the text-critical principle of lectio difficilior recommends the less common reading *toknît*.[189] *toknît* may be based on

---

[187]G. Widengren offered an interpretation of the phrase based on Near Eastern parallels. Widengren believed the term *toknît* was the Israelite equivalent to the Mesopotamian 'tablets of destiny' (*ṭup šīmāti*); accordingly, he saw the figure as the guardian or keeper of the tablets (*The Ascension of the Apostle and the Heavenly Book*. King and Saviour 3. Uppsala: Lundequists, 1950), 26. Although an interesting proposition, this cannot be supported on comparative Semitic linguistic grounds, nor is there any corroborating evidence in the Hebrew Bible. Other suggestions which emend the consonantal text include 'bridegroom of Tanith' (J. Smith, *The Book of the Prophet Ezekiel* [London: SPCK, 1931], 75, reading *ḥôtēn tanît*); 'perfectly wise' (G. A. Cooke, *A Critical and Exegetical Commentary on the Book of Ezekiel* [ICC; Edinburgh: T & T Clark, 1936], reading *ḥākām lətaklît*); 'perfect in form' (P. Cheminant, *Les prophéties d'Ezéchiel contre Tyre (XXVI–XXVIII 19)* [Paris: Letouzey et Ane, 1912], reading *tammîm tabnît*).

[188]The correct reading may be תכונתו 'its arrangement', which appears in 43:11.

[189]The word *tabnît* 'image' is used in Ezek 8:3 with the meaning 'form' or 'image' referring to the hand of God, and again in v. 10, in reference to painted images of animals on the wall of the sanctuary court.

the root *tkn* 'to regulate, measure, estimate'. In Ezek 43:10, it is used as the object of the verb *mdd* 'to measure'.[190] As a result, *toknît* is commonly rendered 'measurement, proportion' or 'example'. Still, this does not lend immediate clarity to the idea behind the phrase *ḥôtēm toknît* in Ezekiel 28. Zimmerli characterized the general semantic range of the verb *tkn* as containing the idea of "correctness," and translated the phrase 'completed signet'. He also, however, refrained from suggesting what the phrase might mean.[191] Similarly, RSV renders *toknît* in 28:12 'perfection'. Both senses are suggestive of conformance to an ideal, and neither is far removed from the root meaning 'to measure'. In fact, 'pattern', 'model', 'regulate', 'measure', and 'estimate' are words within the semantic sphere of conformance to an ideal.

As for the participle *ḥôtēm*, a survey of the objects taken by the verb *ḥtm* demonstrates its semantic range, and three uses seem clear: to stamp or close a document (1 Kgs 21:18; Neh 10:1; Isa 8:16; Jer 32:10, 11; Dan 12:4); to store or hold for safe-keeping (Deut 32:34); to block, cover, or otherwise impede (Job 9:7; Cant 4:12). Other occurrences are questionable with respect to meaning and are accompanied by textual difficulties, suggesting that the ancient scribes were equally at a loss. These various meanings caution against inferring a simple answer. If *toknît* does signify 'measurement, proportion' or 'example', then it would be the object of quite an unusual use of the *verbal* sense of *ḥtm*.

NRSV reflects a widely held decision to accept the reading *ḥôtam* 'seal' (construct form) for *ḥôtēm*, a reading suggested by the Greek translation ἀποσφράγισμα. The central idea is that of a cylinder or stamp seal used by a person to impress his or her "signature" (as it were) on a document or similar item, rendering the item an "official" expression of that person's will.[192] Relevant for our discussion is the

[190] There is some ambiguity with this passage as well; however, in view of the textual variations, the verse throws little direct light on the meaning in Ezek 28.

[191] Zimmerli cites the following as examples reflecting the underlying notion of 'correctness': 'to be right' (*Niphal*) in 18:25, 29; 'to put right' (*Piel*) in Isa 40:13; 'to determine the measure' in Isa 40:12; Job 28:25; and 'to test' (*Qal*) in Prov 16:27; 21:2; 24:12 (*Ezekiel*, see especially 2.81, 91–92).

[192] It is not completely clear whether to understand 'seal' as the signet ring or the impression it leaves. Outside the Septuagint ἀποσφράγισμα regularly signifies the

original significance of the seal, which is thought to be legal, having emerged along with capital formation as a mark of ownership or contractual obligation by an individual and as a symbolic representation of the individual.[193] It may be said that the seal represents or signifies the "essence" of a person.[194]

One example of such usage may be seen in the Mesopotamian phenomenon of votive seals. These seals which bore inscriptions praying for the life of the donor were deposited, just like votive statues, in the sanctuary "to convey that prayer to the deity in place of the donor himself or his statue."[195] That a seal could serve in the same capacity as a statue shows that the ideology of statue and seal are related.

There is a passage which demonstrates this significance with re-

---

impression of a seal, and *signaculum* (the Vulgate rendering) is used in general to convey a "mark" or "sign," and in particular a seal or a signet (See LSJ for examples of αποσφραγισμα signifying the impression of a seal.) In this regard, J. Hermann translated 'stamp of the image' (*Ezechiel, übersetzt und erklärt* [KAT; Leipzig: Keichert, 1924]), J. Ziegler translated 'impression of the original' (*Ezechiel* [Göttingen: Vandenhoeck und Ruprecht, 1952]) and Fohrer rendered 'copy of a model' ("der Abdruck eines Modells"; *Ezechiel* (Tübingen: J. C. B. Mohr [P. Siebeck], 1955.). Zimmerli's statement that "חותם does not contain the idea of reproduction or stamping" (*Ezekiel* 2.81) seems unwarranted; for central to the notion of a seal is the reproduction of a particular image. It might be said that it does not matter whether what was intended was the seal or the seal impression; or better, that the impression itself is not the only aspect in the discussion of likeness. A seal impression is best taken as *evidence* of the seal. In reality each is no more than a mirror image of the other, and that which is represented might be said to lie somewhere *in between* the intaglio of the seal and the cameo of the impression; the representation cannot be simply equated with either, but both bear witness to it.

[193]See W. W. Hallo, "'As a Seal upon Thine Arm': Glyptic Metaphors in the Biblical World," *Ancient Seals and the Bible* (ed. L. Gorelick and E. Williams-Forte; Malibu, Calif.: Undena, 1983), 8. See also J. Renger, "Legal Aspects of Sealing in Ancient Mesopotamia," *Seals and Sealing in the Ancient Near East* (ed. M. Gibson and R. D. Biggs; Bibliotheca Mesopotamica 6; Malibu, Calif.: Undena, 1977), 75–88.

[194]In Genesis 38 Judah's seal is used to identify him. Here, Judah established his identity and legally bound himself to a pledge by leaving his seal. Hallo interprets the cord and 'staff' as the cord on which the seal is worn (around the neck) and the pin with which the seal is connected to the cord (Hallo, "'As a Seal upon Thine Arm'," 8, 14).

[195]Hallo, "'As a Seal upon Thine Arm'," 9. Cf. the Egyptian funerary practice of placing shawabti figurines to do the labor of the deceased.

spect to the royal office. In Gen 41:39–45, Pharaoh makes Joseph his majordomo and vice-regent through the official act of investing Joseph with his seal. In vv. 40–44 he proclaims:

> 40אַתָּה֙ תִּהְיֶ֣ה עַל־בֵּיתִ֔י וְעַל־פִּ֖יךָ יִשַּׁ֣ק כָּל־עַמִּ֑י רַ֥ק הַכִּסֵּ֖א אֶגְדַּ֥ל מִמֶּֽךָּ׃
> 41וַיֹּ֥אמֶר פַּרְעֹ֖ה אֶל־יוֹסֵ֑ף רְאֵה֙ נָתַ֣תִּי אֹֽתְךָ֔ עַ֖ל כָּל־אֶ֥רֶץ מִצְרָֽיִם׃ 42וַיָּ֨סַר
> פַּרְעֹ֤ה אֶת־טַבַּעְתּוֹ֙ מֵעַ֣ל יָד֔וֹ וַיִּתֵּ֥ן אֹתָ֖הּ עַל־יַ֣ד יוֹסֵ֑ף וַיַּלְבֵּ֤שׁ אֹתוֹ֙ בִּגְדֵי־
> שֵׁ֔שׁ וַיָּ֛שֶׂם רְבִ֥ד הַזָּהָ֖ב עַל־צַוָּארֽוֹ׃ 43וַיַּרְכֵּ֣ב אֹת֗וֹ בְּמִרְכֶּ֤בֶת הַמִּשְׁנֶה֙ אֲשֶׁר־
> ל֔וֹ וַיִּקְרְא֥וּ לְפָנָ֖יו אַבְרֵ֑ךְ וְנָת֣וֹן אֹת֔וֹ עַ֖ל כָּל־אֶ֥רֶץ מִצְרָֽיִם׃ 44וַיֹּ֧אמֶר
> פַּרְעֹ֛ה אֶל־יוֹסֵ֖ף אֲנִ֣י פַרְעֹ֑ה וּבִלְעָדֶ֗יךָ לֹֽא־יָרִ֨ים אִ֜ישׁ אֶת־יָד֛וֹ וְאֶת־רַגְל֖וֹ
> בְּכָל־אֶ֥רֶץ מִצְרָֽיִם׃

> 40"You shall be over my house, and all my people shall be obedient[196] to your command; only as regards the throne will I be greater than you." 41And Pharaoh said to Joseph, "See, I have set you over all the land of Egypt." 42Then Pharaoh took his signet ring from his hand and put it on Joseph's hand, and arrayed him in garments of fine linen, and put a gold chain about his neck; 43and he made him ride in his second chariot; and they cried before him, "O vizier![?]"[197] Thus he set him over all the land of Egypt. 44Moreover Pharaoh said to Joseph, "I am Pharaoh, and without your consent no man shall lift up hand or foot in all the land of Egypt."

In this narrative, Joseph becomes, practically speaking, the essence or avatar of Pharaoh as the bearer of Pharaoh's seal. He is the virtual ruler of Egypt.[198] Nevertheless, the people did not give attention to Joseph for Joseph's sake; they in effect gave attention *to the seal* now *in Joseph's possession,* which embodied and represented Pharaoh himself.[199] Through bearing the seal, Joseph has become, for all intents and purposes, Pharaoh — an idea made explicit in the statement, '*I* am

---

196See ***HALOT***, 731, for literature on the meaning of *nāšaq* here.

197On the much-discussed Hebrew word *ʾabrēk*, see most recently E. Zurro Rodríguez, "El hápax *ʾabrēk* (Gn 41,43)," *Estudios Bíblicos* 49 (1991): 269.

198The word for seal here is *ṭabbaᶜat*. Both *ḥôtām* and *ṭabbaᶜat* are most likely Egyptian in origin (see T. O. Lambdin, "Egyptian Loan Words in the Old Testament," *JAOS* 73 [1953]: 145–55).

199By this I do not wish to imply any sense of "mystical" embodiment. The import is ideological and practical. Whoever had been granted the seal presumably would have been treated in such a manner.

Pharaoh, and without *your* consent...'.[200]

The "Promotion of Joseph" narrative helps explain the meaning of two other passages in the Hebrew Bible according to which the seal is used with reference to royal authority. In Hag 2:21–23 the prophet proclaims the impending overthrow of the earth's ruling nations, and the establishment of Yahweh's kingdom. In the new age, Zerubbabel will be elevated to a new status on the world stage: 'On that day, declares Yahweh of hosts, I will take you Zerubbabel my servant, the son of Shealtiel, says Yahweh, and I will make you like the signet (*kaḥôtām*); for I have chosen you, says Yahweh of hosts'. This in all likelihood signifies that Zerubbabel, who is called governor of Judah (2:2,21), will reign in the new kingdom as Yahweh's representative and vice-regent over the nations. Zerubbabel as Yahweh's seal is another expression of the significance of the seal demonstrated in the Joseph narrative.

The notion of a seal symbolizing the king as Yahweh's vice-regent is similarly found in Jer 22:24–25: 'As I live, declares Yahweh, if Coniah the son of Jehoiakim, king of Judah, is a signet (*ḥôtām*) on my right hand, then from there shall I tear you; and I will place you into the hand of those who seek your life...even into the hand of Nebuchadrezzar king of Babylon'. This may refer to a disowning of the Davidic king as Yahweh's representative (cf. v. 30), or it may signify that the very symbol of his presence and authority would be given over for desecration by virtue of its being placed into the hands of a foreign monarch.[201] In either case, it is clear that the exile of Jehoiachin (Coniah) is interpreted as the once vice-regent being lowered to the status of subservience to a foreign ruler.[202] In both cases, the one thing that seems to be quite clear is that the seal is the representative or representation of the owner, in this case Yahweh.[203] The case in Haggai

---

[200]The same situation occurs in Esth 3:10 with regard to Haman, and Esther 8:3 with regard to Mordecai (note especially 8:8). The word in each case is *ṭabbaᶜat*. On presentation seals in Mesopotamia see J. A. Franke, "Presentation Seals of the Ur III/Isin-Larsa Period," *Seals and Sealings in the Ancient Near East*, 61–66.

[201]Cf. Ezek 24:21, Ps 89:40–43. See also, with respect to the sanctuary, Ezek 7:22.

[202]Cf. 2 Kgs 24:12–15.

[203]One other example in which a person is referred to as a seal occurs in the Hebrew Bible: Cant 8:6, 'Place me as the seal upon your heart (*śîmēnî kaḥôtām ᶜal*

is particularly important to this study, for it reveals that the seal as an image of royal authority was in common parlance during the early postexilic period, not long after Ezekiel's writings.[204]

'Seal', 'image', and 'model' convey the ideas of form and conformance. Despite the difficulty in rendering the phrase *ḥôtam toknît* (emended from *ḥôtēm*) in English, it is quite likely that the intended sense lies somewhere within this semantic range. Both G ἀποσφράγισμα ὁμοιώσεος and Vulgate *signaculum similitudinis* may be rendered 'seal of resemblance' and appear to be grammatically literal renditions of the MT, if it were pointed differently. The phrase does not lend easily to English translation; perhaps the best translation is to follow the G and Vulgate 'seal of resemblance'. The phrase may be best understood in view of the Gen 1:26 reference to the creation of humankind in the image (*ṣelem*) and likeness (*dəmût*) of God.[205] Thus, the 'model' expressed in *toknît* or *tabnît* is *ʾĕlōhîm*, in whose image the first human was created. As I discussed in chapter 2, the creation of humankind in the image and likeness of God is a reflection of royal imagery. We have also seen in this section that there is a strong tradi-

---

libbekā); as the seal upon your arm (*kaḥôtām ʿal zərôʿekā*)'. It is difficult to discern the precise meaning of this verse; still, the words that follow ('for love is as strong as death; jealousy is as harsh as the grave') make clear that it is to be regarded a statement of considerable force. On the interpretation of this verse, see M. Pope, *Song of Songs* (AB; Garden City, N.Y.: Doubleday, 1964), 666–69. Note the interpretation of Hallo, that the beloved seeks "physical intimacy" and "the status-symbol seal function assumed by the seal in burials: even as the latter perpetuates the owner's standing in death, so the beloved declares that her lover's role with regard to her will outlive him." W. W. Hallo, "'As a Seal upon Thine Arm'," 12–13.

[204]Note also the examples with *ṭabbaʿat* in Esther 3:10 and 8:3.

[205]This idea may be related to the much later Christological statements of the New Testament, which are widely recognized as deriving from the language of first man traditions. Heb 1:3 refers to Christ as the 'stamp of the nature of God' χαρακτηρ τῆς υποστάσεως, the word χαρακτηρ meaning a seal impression. This is reiterated in similar terms in 2 Cor 4:4 which calls him the εἰκων τοῦ θεοῦ 'image of God' — revisiting the phrase κατ' εἰκόνα of Gen 1:26. What is particularly interesting here is that the phrase εἰκων τοῦ θεοῦ is applied to Christ as a mark of distinction — a sense different from the context of Genesis, which suggests that it is democratized to imply all humans. One would hardly think this statement was borrowed directly from Ezekiel. It is more likely that it belonged to the pool of traditions and images about the first man that was drawn from for a variety of applications.

tion in the ancient Near East associating the seal with royal imagery (although, of course, nonroyals can have and use seals). Although translation is difficult, it seems safe to consider the phrase a term suggesting royal authority.

In the cases of Coniah (Jehoiachin) and Zerubbabel, the seal imagery was an expression of their role as Yahweh's vice-regent. In Mesopotamia, the practice of kings considering themselves the vice-regent of a god is well known. In Assyrian royal inscriptions, the kings referred to themselves with terms such as 'governor of Enlil' (*šakin* *ᵈEnlil*) and 'vice-regent of Aššur' (*iššâk* *ᵈAššur*).[206] It is significant for our purposes to note the observation that the epithet "vice-regent of Aššur" alluded specifically to the king as "intermediary between the god and the community."[207] In this respect, Coniah and Zerubbabel were expected to execute the divine will. As far as the primal human goes, it is enough to note that the language points to and underscores his position between God and humanity.

Beauty: a royal attribute (12b)

It is worth asking whether the reference 'perfect in beauty' (*kəlîl yōpî*) in v. 12 carries any technical significance, which may cast light on the identity of the figure, that is, whether it can in some way be considered more than an incidental literary embellishment. Once again, I believe there is warrant for interpreting the allusion as having a background in royal ideology. The imagery of "beauty" is a common trope in the Hebrew Bible. It is frequently used with reference to women, and in this sense strongly suggests the idea of physical beauty as an attractive force. This is particularly evident when it appears elsewhere in Ezekiel, where it metaphorically describes Jerusalem as a woman (16:14, 15, 25). The idea is used similarly of Zion in Ps 50:2, where it is used in combination with a nominal derivative of the root *kll*, as is likewise true of our passage. In fact, this combination appears to be

[206]See P. Machinist, "Literature as Politics: The Tukulti-Ninurta Epic and the Bible," *CBQ* 38 (1976): 465 and references.

[207]M. T. Larsen, *The Old Assyrian City State and Its Colonies* (Copenhagen: Akademisk Forlag, 1976), 119.

rather common, appearing elsewhere in Lamentations as *kəlîlat yōpî* (2:15), a phrase shared by Ezekiel in 27:3.

On the mythological level, the reference to beauty may be explained by reference to language applied to the king. Isa 33:17 prophesies, 'Your eyes will see the king in his beauty' (*melek bəyōpyô*). In Mesopotamia an account of the creation of humankind and the creation of the king is found in text VAT 17019, as discussed above in chapter 2. The description of the creation of the king is given as follows:

> *ṭa-a-bi u[b]-bi-ḫi gi-[mi]r la-a-ni-šú*
> *ṣu-ub-bi-i zi-i-mi-šú bu-un-ni-i zu-mur-šú*
> ^{d}*Be-let*-DINGIR.MEŠ *ip-ta-ti-iq* LUGAL *ma-li-ku* LÚ
> *id-*⌜*di*⌝*-nu-ma a-na* LUGAL *ta-ḫa-za* DINGIR.ME[Š GAL.MEŠ]
> ^{d}*A-nù it-ta-din a-gu-ú*(Text: PA)*-šú* ^{d}EN.LÍL *it-[ta-din kussâ-šú]*
> ^{d}U.GUR *it-ta-din* ⌜^{giš}TUKUL⌝.MEŠ*-šú* ^{d}MAŠ *i[t-ta-din šá-lum-mat-su]*
> ^{d}*Be-let*-DINGIR.MEŠ *it-ta-din bu-un[-na-ni-šú]*
> *ú-ma-ʾi-ir* ^{d}*Nusku ú-ma-lik-ma iz-z[iz ma-ḫar-šu]*[208]
>
> "With goodness envelop his entire form.
> Form his *features harmoniously*; make his *body beautiful*!"
> Thus did Belet-ili construct the king, the counselor-man.
> The great gods gave the king *warfare*.
> Anu gave him his *crown*; Enlil ga[ve him his *throne*.]
> Nergal gave him his *weapons*; Ninurta ga[ve him his glistening *splendor*.]
> Belet-ili gave him [his] *beautifu[l appearance*.]
> Nusku gave instruction, he taught counsel and stoo[d by him] as a servant.[209]

This passage is reminiscent of Psalm 45:3–4, which similarly lauds the comeliness and "divine attributes" of the king:

3 יָפְיָפִיתָ מִבְּנֵי אָדָם
הוּצַק חֵן בְּשְׂפְתוֹתֶיךָ
עַל־כֵּן בֵּרַכְךָ אֱלֹהִים לְעוֹלָם׃
4 חֲגוֹר־חַרְבְּךָ עַל־יָרֵךְ גִּבּוֹר
הוֹדְךָ וַהֲדָרֶךָ׃

---

[208]Text from W. Mayer, "Ein Mythos von der Erschaffung des Menschen und des Königs," *Or* 56 (1987): 55–68, see 56.

[209] Lines 34–41. Italics here indicate emphasis.

3 You are the fairest of the sons of men;
grace is poured upon your lips;
therefore God has blessed you for ever.
4 Gird your sword upon your thigh, O mighty one,
in your glory and majesty.[210]

The latter lines of VAT 17019, particularly the reference to Anu bestowing a crown and Ninurta bestowing splendor, recall Ps 8:6:

וַתְּחַסְּרֵהוּ מְּעַט מֵאֱלֹהִים
וְכָבוֹד וְהָדָר תְּעַטְּרֵהוּ

You have made him a little lower than *ʾĕlôhîm*
You have crowned him with glory and honor.

The attention given to the physical characteristics of the king is also found in the Gilgamesh Epic. In the introduction of the epic, Gilgamesh is described in the following terms (I 43–50):

[*šá* / *mannu it-*]⌜*ti*⌝-*šu iš-šá-an-na-nu a-na šarrūti*(LUGAL^{ti})
[*šá kīma*(GIM)] ^{d}GIŠ.GÍN.MAŠ *i-qab-bu-ú a-na-ku-ma šarru*(LUGAL)
[^{d}GIŠ.GÍ]N.MAŠ *iš-tu* u_4*-um i*ʾ*-al-du na-bu šum-šú*
šit-tin-šú ilu(DINGIR)-*ma šul-lul-ta-šú a-me-lu-tu*
ṣa-lam pag-ri-šu DINGIR.MAḪ *uṣ?-ṣi*[*r?*]
[*uš* / *l*]-⌜*te*⌝-*eṣ-bi*-⌜*i*⌝ *gat-ta-šú u-ṭàr-ra* [*x*]
[x x x x (x) *da*]*mqu* ([S]IG$_5$) *šá-ru-uḫ eṭl*[*ūti*] (GURUŠ.[MEŠ])
[x x x x x x x] *gít-ma-l*[*u?* x x x x (x)][211]

[Who(?)] can be compared with him in kingship,
[Who like] Gilgamesh can say, "I am king indeed!"?
[Gilg]amesh was summoned by name from the very day of his birth
Two-thirds of him is god, one-third of him is human.
The image of his body Dingirmaḫ desig[ned(?)].
....[com]pleted his form (...?) like(?)....
....[fa]ir, most glorious [among heroes].
....perfect...

---

[210]The verbal form יְפְיָפִיתָ derives from יפה, but is completely anomalous. It resembles the *pəᶜalᶜal* conjugation but is considered by most a scribal error. Read perhaps יָפִיתָ (GKC §55e), or יָפוֹ יָפִיתָ. On the connection of Ezek 28:12 and Psalm 45 to royal ideology of the ancient Near East see K. Yaron, who also adds the description of Pharaoh in Ezek 31:7–9 ("The Dirge over the King of Tyre," *ASTI* 3 [1964]: 47).

[211]For text and translation, see J. Tigay, *Evolution of the Gilgamesh Epic*, 142, 264.

Unfortunately, the broken text precludes precise interpretation. Nevertheless, the beauty of the king here, as elsewhere, seems to be related to his similarity to the gods.

Eden, the 'garden of God' and the holy mountain of God (13–14)

The phrase *bəᶜēden gan ᵓĕlōhîm hāyîtā* in v. 13 is obviously reminiscent of the Genesis 2–3 narrative, which we discussed in the last chapter. As we saw there, Eden represents the divine abode. Just as in the first oracle of Ezek 28:1–10, so here the protagonist is placed in the divine habitation.[212] It was referred to as the seat or dwelling of the gods in 28:2, and here it is referred to as Eden, the garden of God. The phrase may also be translated 'garden of the gods' or 'divine garden'. But this location is also called *har qōdeš ᵓĕlôhîm* 'the holy mountain of God' in v. 14 and *har ᵓĕlôhîm* 'the mountain of God' in v. 16; the garden and the mountain are equated as the divine habitation.[213]

There is, no doubt, significance in the fact that the protagonist is stated to be in the divine habitation: the protagonist was a human living in a divine place — a situation identical to that presented in the first oracle in 28:1–10. I have already discussed the connection of the primal human and the garden of God as an image of kingship. Although we do not find that motif here, we are told something of the significance of Ezekiel's conception of the primal human in this setting in the verses that follow, to which we shall now turn.

The endowment of the powers of divination (13)

The text continues in v. 13 *kol-ᵓeben yəqārâ məsūkātekā* 'every precious stone was your *məsūkâ*'. Given the context of royal imagery, the list of stones that follows, and extrabiblical evidence which I shall cite below, it becomes clear that the primal human figure is endowed with a sacral investiture. Once again, the language is obscure. In particular, there is considerable ambiguity about the noun *məsūkâ,*

---

[212]For discussion of Ezek 28:1–10, see chapter 5.

[213]See discussion above in chapter 2.

which renders all interpretations tentative. It may be understood to derive from the root *swk* 'to hedge, fence in'. The verb appears in the well-known passage of Job 3:23, which speaks of the protection or shielding from adversity which the accuser cites as the reason for Job's remarkable piety. The noun *məsûkâ* appears in Mic 7:4 with the meaning 'hedge', although the use is figurative and the precise meaning is uncertain. The notion of a hedge or shield here in Ezekiel makes some sense when viewed alongside the protective cherub of v. 14. But if one accepts the sense of "protection," it is not clear why a being on the mountain of God would require protection (and from what). What is more, one would have to go to great lengths to corroborate the notion of "protective" stones in the ancient Near East.

məsūkâ may, alternatively, be understood to derive from the root *skk* 'to overshadow, screen', despite the fact that the *k* is not doubled.[214] There are two other nouns related to this root, *māsāk*, generally translated 'covering, screen', and *mûsāk*, an apparent technical term in architecture. The context of *māsāk*, the more frequent of the two, suggests something that cannot be seen through. It appears frequently in descriptions of the tabernacle, signifying the multicolored cloth screen at the gate of the court.[215] The word *mûsāk* appears only once, in 2 Kgs 16:18, and refers to some type of cultic structure, which RSV translates as 'covered way'.[216] It is likely that *məsūkâ* is a technical term from the temple cult, but the lack of external evidence precludes a more precise definition. It is reasonable to understand precious stones as connected with some type of cultic article that is perhaps worn (hence the translation 'covering'). The agreement of the

---

[214]Middle weak and geminate roots are often variants of each other.

[215]The word is also used in a secular context in 2 Sam 17:19, where it refers to something spread over the mouth of a well. There also appears to be a connection to the idea of protection. In Ps 105:39 *māsāk* is used with reference to the cloud of the Exodus wilderness narratives, and draws attention to the protective quality of the cloud in blocking the rays of the sun. Similarly, in Isa 22:8 it suggests the idea of protection, although it is not immediately clear whether the word refers to a specific item or it simply means 'covering' in the unspecified generic sense.

[216]Textual problems add to the difficulty in identifying the intended item.

list of stones with the stones of the sacred breastplate naturally suggests the breastplate as the intended referent, but this must be examined more closely.

The list of precious stones in Ezekiel 28 may be a simple literary depiction of an idea such as the wealth of Tyre, but it is more likely that the names of the stones are given to signify and draw attention to another referent. The names of the nine stones associated with the *məsūkâ* immediately recall the twelve stones of the high priestly pectoral in Exod 28:17–20, although there is some rearranging of the order. Several of the stones appear individually elsewhere in the Hebrew Bible, outside the lists of Ezekiel 28 and Exodus 28, but it is difficult to attach any significance to them as individual items; the significance, I believe, lies in the list as a totality. Our attention is immediately directed to the high priestly pectoral as the context for the list, but also to the fact that the list in Ezekiel 28 is incomplete.

MT Ezekiel 28	MT Exodus 28	LXX Ezekiel 28	LXX Exodus 28
(1) ʾōdem	*(1)ʾōdem*	*sardion*	*sardion*
(2) piṭdâ	*(2) piṭdâ*	*topazion*	*topazion*
(3) yahălōm	*(9) bāreket*	*smaragdos*	*smaragdos*
(4) taršîš	*(8) nōpēk*	*anthrax*	*anthrax*
(5) šōham	*(7) sappîr*	*sappheiros*	*sappheiros*
(6) yošpēh	*(3) yahălōm*	*iaspis*	*iaspis*
(7) sappîr	*(*) lešem*	*ligurion*	*ligurion*
(8) nōpēk	*(*) šəbû*	*achatēs*	*achatēs*
(9) bārkat	*(*) ʾaḥlāmā*	*amethyston*	*amethyston*
	(4) taršîš	*chrysolithos*	*chrysolithos*
	(5) šōham	*bēryllion*	*bēryllion*
	(6) yošpēh	*onychion*	*onychion*

The Greek text of Ezekiel 28 text includes the names of twelve stones, not the nine of the MT, and the list matches the Greek text of the Exodus passage. Like MT, the Vulgate (not pictured) includes nine stones, which appear in the same order, with the exception of the last three. The Syriac (not pictured), on the other hand, has a list of eight, three of which correspond to the Exodus list. When the list of MT

Exodus 28 is placed alongside the list of MT Ezekiel 28, it becomes apparent that the overall enumeration of the list is different, but the groups remain essentially the same. The four differences to note are: a) stone group *1-2-3* is broken up in Exodus 28 by stone group *7-8-9*; b) stone group *7-8-9* appears in Exodus 28 in reverse order as *9-8-7*; c) three stones in Exodus 28 which do not appear in Ezekiel's list follow the first two groups; and d) the list in Exodus 28 ends with stone group *4-5-6*. The similarity of the two lists makes it obvious that the enumeration of stones in Ezekiel 28 is set up to recall the high priestly pectoral. On the other hand, the differences are significant enough to rule out a simple case of literary borrowing.[217] Thus, it is best to conclude that the two texts are based on an earlier common tradition. It is possible that the list is not original to the Ezekiel oracle, but the fact that it appears in multiple witnesses recommends its antiquity.[218] At whatever stage it was inserted, the original intent was, in all likelihood, to preserve the traditional understanding of the phrase *kol-ʾeben yəqārâ*.

U. Cassuto has suggested that the stones of the breastpiece intentionally recalled the precious stones of the garden:

> The precious stones which are mentioned in Ex. 28, 17–20 are the same precious stones which, according to the Ezekiel tradition in 28:13, were found in Eden, the garden of God. The Pentateuch did not take into its story of the garden of Eden the precious stones and the gold out of opposition to this tale, but says that they were found outside the garden (Gen 2:11–12). However, the Pentateuch did use the stones as a symbol in popular tradition: the high priest absolving the sins of the people carries on his chest the gold and the precious stones, which are a symbol of the situation in Eden, when man was still innocent.[219]

This hypothesis is impossible to prove, but it is attractive in that it suggests the interaction of these traditions.

---

[217]According to Yaron, the LXX preserves the original list, based on its agreement with Exod 28; the list was subsequently corrupted in MT (K. Yaron, "The Dirge over the King of Tyre," especially 35–36).

[218]Zimmerli elected to omit the list of stones from the Hebrew text, and considered it a later gloss on the basis of what he had identified as parallelism between the phrases that precede and follow the list. ***Ezekiel***, 2.82.

[219]Cited by Yaron, "The Dirge over the King of Tyre," 44–45.

There is a later tradition which associates the cherub with the stones of the high priestly breastpiece. Pseudo-Philo preserves a tradition according to which the stones were placed on the cherubim of the temple. In his retelling of the Joshua story, the author of Pseudo-Philo tells of twelve precious stones obtained from the Amorites. According to the tradition he cites, these stones came from the land of Havilah — a reflection of Gen 2:11–12, which relates that Havilah has gold and precious stones.[220] These stones were of a supernatural nature, and were set into seven golden idols known for their divinatory powers, housed in the temples of the Amorites.[221] The stones were destroyed by God because they were defiled by the Amorites; and God himself later replaced them with twelve new stones taken from the same place (Havilah). God then commanded that they be kept with the ephod, and later:

> God said to Kenaz, "Take those stones and put them in the ark of the covenant of Yahweh along with the tablets of the covenant that I gave to Moses on Horeb; and they will stay there until Jahel, who will build a house in my name, will arise, and then he will set them before me upon the two cherubim, and they will be before me as a memorial for the house of Israel.[222]

There is no evidence of this practice in the Hebrew Bible, of course, but neither is there reference to where the breastpiece was kept when it was not being worn. Pseudo-Philo's narrative suggests the cultic and divinatory background of the stones in Ezekiel, which accords nicely with the argument below that the explanation of the appearance of the stone list in Ezekiel is to be found, in part, in the earlier phrase *mālēʾ ḥokmâ* 'full of wisdom'.

The reference to the stones of the breastpiece reveals the priestly orientation of Ezekiel's conception of the primal human. The signifi-

---

[220]Of the two stones cited in Gen 2:12, one, *šōham*, is found in the lists of Ezekiel 28 and Exodus 28.

[221]According to the text, these idols, 'when called upon, showed the Amorites what to do every hour' (25:11). Cf. 25:12. Translation from D. J. Harrington in *The Old Testament Pseudepigrapha*, ed. J. H. Charlesworth (New York: Doubleday, 1985).

[222]Pseudo-Philo 26:12.

cance of the primal human bearing what amounts to the breastpiece of the high priest is cultic, and an expression of divination. The breastpiece of judgment contained the "urim" and "thummim," the sacred lots, which were the means by which the will of the deity was made known. What this text describes, then, is the primal human's access to divine knowledge. The imagery is consonant with that of the endowing of the king with divine attributes, a topos we have mentioned above. Just as the king (or hero) was endowed with physical beauty which expressed his close association with the gods, he was also endowed with wisdom or access to divine knowledge. In an inscription of Ashurbanipal, the king describes himself as someone:

> [For whom Ashur], father of the gods, decreed a kingly destiny while I was still in my mother's womb,
> [Whose name Nin]lil, the great mother (of the gods), named for the rulership of the land and people,
> [....whose figure] Dingirmaḫ made into the image of a lord...;
> [?] Sin, the holy god, *caused me to see good omen*[s] that I might exercise sovereignty;
> [Shamash and Adad] *entrusted to me the never-failing craft of divination;*
> [Mar]duk, sage of the gods, *gave me a gift of great intelligence and broad understanding;*
> Nabu, the universal scribe *made me a present of the precepts of his wisdom.*[223]

Further accouterments: timbrel and pipes (13)

The phrase in v. 13 (MT) ***məleʾket tuppeykā ûnəqābeykā bāk*** is difficult. I believe it is, in part, a corruption of the textual tradition that is reflected in MT Exod 28:17–20. As we have just seen, the stone list recalls the list of Exod 28:17–20, which describes the high priestly breastpiece. At the end of the list we read (v. 20), ***zāhāb yihyû bəmillûʾōtām*** 'They shall be of gold in their settings'. After the final stone in our text in Ezekiel (***bārkat*** in v. 13, cf. Exod 28:17 ***bāreket*** and the above table) the text continues, ***wəzāhāb məleʾket***. Placing the two texts side by side reveals that the words מלואתם and מלאכת are enough alike that one may be the result of a textual corruption,

---

[223]Ashurbanipal L^{4} (emphasis mine); see translation and references in J. Tigay, *Evolution of the Gilgamesh Epic*, 154.

although no simple text-critical explanation is apparent. Because Exod 28:20 reads *bəmillûʾōtām* with *zāhāb*, I believe *məleʾket* in Ezek 28:13 is best read with *wəzāhāb* and the preceding stone list (ignoring the athnach), and not with *tuppeykā* and *nəqābeykā*. Therefore, I have emended the text and offered for *wəzāhāb məleʾket* the provisional translation 'and of gold was (its) workmanship'.[224]

I believe it is best, however, to retain the MT reading נקביך and translate 'your pipes'. First, there is a dearth of evidence to assign the word *tuppeykâ* a meaning other than 'your timbrels'. Of the 17 times the word *tōp* occurs in the Hebrew Bible, it is conventionally rendered 'timbrel', and in every case but one (Ezekiel 28), the meaning is clear and, for the most part, uncontested.[225] Second, the translation 'your pipes' for *nəqābeykā* can be justified on linguistic and textual grounds.

The normal word for 'flute' in the Hebrew Bible is *ḥālîl*, a word semantically generated from the root meaning 'to bore'.[226] The word

---

[224]Note the emendations in the above notes on the text. The construct form precludes reading *məleʾket* with anything other than what follows. Many commentators have sought to overcome the problems in the text by abandoning the Masoretic division and reading *zāhāb* with what follows. The Greek text has provided the main impetus for this departure from MT, reading και χρυσιου ενεπλησας τους θησαυρους σου και τας αποθηκας σου εν σοι 'and with gold you filled your treasuries and stores in you'. In the last century Cornill translated 'und aus Gold war gearbeitet deine Fassung(?) und deine Vertiefungen(?)'. (C. H. Cornill, *Das Buch des Propheten Ezechiel* [Leipzig: J. C. Hinrichs, 1886], 360–61). Likewise G. A. Cooke, *Ezekiel*, 316; Fohrer, *Ezekiel*, 161. BDB understood this passage to express the idea of service or use, and cited Lev 7:24; 11:32; and Judg 16:11 as examples of this sense. Each of the three verses uses a form of the verb *ʿśh*. Lev 11:32 and Judg 16:11 include the preposition *b-*, and for this reason, seem to resemble the example in Ezek 28:13. They are, nonetheless, quite different from a grammatical standpoint and are therefore of little value in clarifying our verse, since in both cases, the *b-* is generated by the need for a resumptive pronoun to complete a relative clause.

[225]Zimmerli bristles at the notion of relating the word to timbrel, stating that it is "scarcely to be considered"; he, however, offers no reason for this dismissal. He does offer the caveat that the meaning of the word "remains completely obscure" (*Ezekiel*, 2.84).

[226]BDB classifies this, rightly so, under the root *ḥll* (I) 'bore, pierce'. There is ample evidence elsewhere in Semitic. Cf. Ethiopic *ḫel(l)at* '(hollow) stick' (W. Leslau, *Ethiopic and South Arabic Contributions to the Hebrew Lexicon* [Berkeley: University of California Press, 1958], 20) and Tigre *ḫel* 'pipe, flute' (E. Littmann and M. Höfner, *Wörterbuch der Tigre-Sprache* [Wiesbaden: Franz Steiner, 1962], 52b). For

nəqābeykā is most reasonably taken from the root *nqb*, which in Semitic carries the sense 'to dig', 'to tunnel', or 'to pierce'.[227] It may be accepted as deriving phonologically from the noun **neqeb*, which conforms phonetically to the noun pattern **qitl* — a noun pattern which often expresses the result of a transitive verbal action.[228] Thus, **neqeb* can, like *ḥālîl*, denote a drilled thing or 'flute, pipe'.

Textual evidence in support of the translation 'pipes' for *nəqābeykā* includes passages which place the articles 'timbrel' (*tōp*) and 'flute' (*ḥālîl*) together. In the reference to ecstatic prophets in 1 Sam 10:5, among their accessories we find *tōp wəḥālîl* 'timbrel and flute' in the context of a list of musical instruments. In Isa 5:12 the two are paired and set over against lyre and harp. It is, then, very reasonable to understand **neqeb/*nēqāb/*nāqāb* as a synonym for *ḥālîl*.

We must next ask the significance of timbrels and pipes in this setting. We have observed in this passage that *nəqābeykā* is paired with *tuppeykā*, and should thus be considered closely related to it. Since *tōp* means timbrel, *nəqābîm* could, therefore, be musical instru ments or celebratory accessories. A plausible translation for the assumed plural *nəqābîm* would perhaps be 'pipes', capturing the semantic image of 'drilled things' or 'things with holes'. The obvious reason that scholars have not considered this translation is that they have not seen the notion of "jubilation" as a part of the primal human myth. There is, however, considerable evidence to suggest otherwise.

Such an interpretation is consistent with other biblical passages. The first human in Job 15:7–8 is spoken of within the context of the

---

references see KB. Note also its noun pattern **qatîl*, which characteristically creates passives from active roots (e.g., *kālîl* 'completed'; *śākîr* 'hired').

[227] Syriac *neqbā*, Mandaic *niqba* 'hole' (see E. Drower and R. Macuch, *A Mandaic Dictionary* [Oxford: Clarendon, 1963], 299b); Arabic *naqb* 'passage, tunnel, narrow pass'; Akkadian *naqbu*; Siloam Tunnel Inscription *nqb(h)*, see Cross and Freedman, *Early Hebrew Orthography* (New Haven: American Oriental Society, 1952), 50.

[228] Although *qitl* is one of several possible derivations (*nəqābeykā* may also derive from **nāqāb* or **nēqāb)*, note the above cited *qvtl* patterns of *nqb* in Syriac, Mandaic, Arabic, and Akkadian. Examples of *qitl* nouns which denote the result of the verbal action are *šēmaᶜ* 'report (what is heard)', *zebaḥ* 'sacrifice (what is offered)', *ḥēleq* 'portion (division)', *zēker* 'memory'. These and other examples are cited by T. O. Lambdin and J. Huehnergard, *Historical Grammar of Biblical Hebrew: Outline* (1985, unpublished manuscript).

divine council; and in 38:4–7, Yahweh alludes to rejoicing among the sons of God at the creation. Similarly, in Prov 8:31, the figure Wisdom, which is related to the notion of the primal human, rejoices at the creation.[229] In the light of these references, it is by no means unreasonable to see this figure in Ezekiel associated with festivity, though it is not immediately apparent how this relates to Tyre.

A highly suggestive extrabiblical parallel is found in the Indo-Iranian traditions. Yama, the primal human in the Rig Veda, drinks under a leafy tree together with the gods; there songs and flute-playing resounds. One hymn says in reference to this:

> This is the dwelling place of Yama,
> that is called the home of the gods.
> This is his reed-pipe that is blown,
> and he is the one who is adorned with songs.[230]

Two examples from Mesopotamia may be cited, both of which place drums and pipes together in a temple setting and suggest cultic significance. The first is a painting on a vessel found at Ashur of an Assyrian offering scene.[231] It depicts a cultic scene featuring a stylized palm tree and a flaming incense stand. To the left of the scene are two musicians; one plays the hand drum, the other the double flute. On a fragmentary ivory plaque from Nimrud, three musicians (of a group) advance in a procession to worship an enthroned goddess. The first plays the double flute; the second plays the hand drum (a third plays a stringed instrument).[232]

One could also cite a text in the Hebrew Bible, which might reflect a cultic practice based on that tradition. Isa 30:29 reads: 'You shall have a song as in the night when a holy feast is kept, and gladness of heart, as when one sets out to the sound of the flute to go upon the mountain (*lābôʾ bəhar*) of Yahweh'. Here, 'the sound of the flute' (*ḥālîl*) is associated with being on the 'mountain of Yahweh'. Isaiah alludes to this image in order to illustrate what he means by 'gladness

---

[229]See my discussion of these passages in chapters 5 and 7, respectively.

[230]10:135. See W. D. O'Flaherty, *The Rig Veda* (London: Penguin, 1981), 56.

[231]Pritchard, *ANEP*, 628.

[232]Pritchard, *ANEP*, 203.

of heart'. The imagery gives rise to, or expresses (or is *synonymous* with) "gladness." It suggests a background sufficient to make an automatic association between pipes and the mountain of Yahweh.

We note in closing that Ezek 28:13 relates concerning the timbrels and pipes that *bəyôm hibbāraʾăkā kônānû* 'in the day you were created, they were established'. The significance of this phrase lies in the common ancient Near Eastern topos of the endowing of the king with divine attributes — an act that takes place at the birth or creation of the king.

The anointed cherub (14)

There are two traditions present in 28:14 with regard to the identity of the cherub. The consonantal text is ambiguous regarding the relationship between the primal human and the cherub. The Masoretic pointing suggests quite clearly that the cherub and the first human figure are one and the same. The Greek and Syriac suggest that the phrase *ʾatt kərûb* 'you were the cherub' should probably be read *ʾet-kərûb* '(you were) with the cherub'.[233] What is more, the tradition preserved in Gen 3:24 seems quite clearly to make a distinction between human and cherub. Because of the Greek and Syriac witnesses and Gen 3:24, I have adopted the reading 'I placed you *with* the anointed cherub'.[234]

The precise meaning of *mimšaḥ* is difficult to discern. It occurs only once in the Hebrew Bible and presumably derives from the root *mšḥ* 'to anoint'.[235] The ancient textual witnesses present divergent

---

[233]It is tempting to suppose a connection between the next word ***hasôkēk*** and *məsūkāteka* of v. 13, both of which belong to the root ***skk***. Perhaps the item called *məsūkāteka* represents some sort of shield. If so, one might then argue for *ʾatt* as representing the second person pronoun (emending for gender); what is presumably the function in ***hassôkēk*** ('who blocks' or 'guards') could then be seen as adumbrated in the attention given to the *məsūkâ*.

[234] The Masoretic understanding of the consonantal text is curious and invites further study, for which space here does not allow.

[235]Considering the construction of the noun as bearing an *m*-preformative does not offer help. Although many Semitic nouns constructed with *m*-preformative denote the idea of place or time (especially Akkadian, Arabic, and Aramaic), this cannot

readings. It is understood to be related to the root idea 'to anoint' by Theodotion in the translation κεχρισμενου. The translation of the Vulgate, *extentus*, reflects an understanding which derives meaning from a notion seen perhaps in the Aramaic and Syriac root *mšḥ* 'to measure out'. Another suggestion has been to relate *mimšaḥ* to the idea 'to glitter, shine' based on the Akkadian words *mišḫu, nimšaḫu*.[236] The lack of a convincing comparative linguistic alternative warrants retaining MT and the provisional translation 'anointed'. *mimšaḥ* may be construed as a governed noun in a construct relation (with *kərûb*), used adjectivally; that is, representing an attribute of the governing noun.[237]

The participle *hassôkēk* in v. 14 is best taken, from a grammatical standpoint, as a noun in apposition[238] 'who guards' or 'who covers', based on other occurrences of the root *skk* in the Hebrew Bible.[239] The context of the oracle does not clearly commend either sense. In no other known text is the root *skk* associated with cherubim; hence, any explanation is conjectural. Several commentators have elected to follow the Greek which lacks an equivalent.[240] Additional support for this translation may be found in Nah 2:6 in which *sōkēk* is used to express a mantelet.

Despite the many references to cherubim in the Hebrew Bible,

---

be considered a reliable guide in Hebrew, which shows a much greater semantic range in this class of nouns.

[236]Tur-Sinai [Torczyner], "Presidential Address," *JPOS* 16 (1936): 5. *AHw*, 623b, 660b; see also KB, 564.

[237]GKC §128p. It is interesting that the phrase *kərûb hassôkēk* recurs in v. 16 without *mimšaḥ*. This draws attention in v. 14 to the omission of the article before *kərûb*, as noted by Zimmerli, who suggests the possibility that it is a name. He nonetheless translates 'the guardian cherub' (*Ezekiel*, 2.85).

[238]It is possible to consider it the final member of the construct chain *kərûb mimšaḥ hassôkēk*. I have translated 'anointed guardian cherub' for convenience.

[239]In Nah 2:6 *sōkēk* signifies a protective apparatus used in warfare. Cooke (*Ezekiel*, 318) saw a link between the idea "cover" in the word's root *skk* (here and v. 16) and the concept of the cherubim of the ark. Recognizing both senses, Zimmerli translated 'guardian', but adds "the word recalls the description of the temple cherubs over the ark." *Ezekiel* 2.85. See also G. Widengren, *Early Hebrew Myths and Their Interpretation*, 166; and idem, *The King and the Tree of Life*, 56.

[240]Cooke, *Ezekiel*, 317; Fohrer, *Ezechiel*, 161; Cornill, *Ezechiel*, 360.

surprisingly little is known about them.[241] The etymology of *kərûb* is unknown.[242] Of one thing we may be certain: cherubim are closely associated with Yahweh and, in a sense, signify the divine presence.[243] It is important to notice that cherubim are closely associated with the king as well. One example is a representation from 10th-century Byblos depicting the throne of King Aḥiram supported by two cherubim.[244] The reference to the cherub, then, in Ezek 28:14 is best understood as a significant statement about the primal human and not a thoughtless embellishment.

The primal human walks amid 'stones of fire' (14)

Another of the more puzzling elements of the text is the reference in v. 14 to the *ʾabnê ʾēš* 'stones of fire'. Outside of our passage, the Hebrew phrase is otherwise unattested. There are two interpretations which predominate: (1) the stones of fire refer to jewels on the mountain of God; (2) the stones of fire refer to divine beings. Evidence in favor of interpreting the stones of fire as jewels may be the jewels on the mountain called Mashu, referred to in the ninth tablet of the Gilgamesh epic (SB version):

> NA$_4$.GUG *na-šá-at i-ni-ib-šá*
> *is-ḫu-un-na-tum ul-lu-la-at a-na da-ga-la* DU$_{10}$*-bat*
> NA$_4$.ZA.GÌN *na-ši ḫa-as-ḫal-ta*
> *in-ba na-ši-ma a-na a-ma-ri ṣa-a-a-aḫ*

---

[241]M. Haran, *Temples and Temple Service in Ancient Israel* (Winona Lake, Ind.: Eisenbrauns, 1978), especially 254–59. See also M. Mach, *Entwicklungsstadien des jüdischen Engelglaubens in vorrabbinischer Zeit* (Tübingen: J. C. B. Mohr, 1992) *passim*.

[242]BDB conjectures a relationship to Akkadian *karābu* 'to be gracious to, bless' cf. the adj. *karūbu* 'honored person' (*CAD*, K, 240). Some have noted the equation *kirūbu* with *šēdu*, the winged bull in Assyria. Earlier it was thought that *kərûb* was related to Greek γρυψ, from Persian *giriften*.

[243]See Isa 37:16. Note especially the copious references to cherubim and the manifestation of Yahweh in Ezekiel (chapters 9–11, 41).

[244]*ANEP*, 456, 458. Note that May (like others) associates the cherub with kingship, particularly with respect to the enthroned king. H. G. May, "The King in the Garden of Eden," in *Israel's Prophetic Heritage* (ed. B. W. Anderson and W. Harrelson; New York: Harper, 1962), 172.

> A carnelian tree was in fruit,
> hung with bunches of grapes, lovely to look on.
> A lapis lazuli tree bore foliage,
> in full fruit and gorgeous to gaze on.[245]

I do not find this reference definitive for our purposes in Ezekiel. The interplay of fruit and precious stones speaks more to the issue of the fertility of the divine mountain.[246] The fact that in the present text of Ezekiel 28 'precious stones' have already been explicitly referred to (v. 13) makes it unlikely in my judgment that we have, once again, a reference to jewels.

The more likely interpretation of the stones of fire is that they refer to divine beings. Some scholars have offered this interpretation by means of emending the text. Thus, Dussaud suggested in place of *ʾabnê ʾēš*, the reading *bənê ʾēl* 'sons of God'.[247] Another common emendation is 'sons of fire'.[248] May has suggested that both interpretations imply a reference to the stars, and with it, the image projected in Job 38:4–7, where it is implied (as we have already discussed in chapter one) that the primal human sang and shouted for joy with the heavenly host, called there both stars and sons of God.[249] I believe that it is possible to maintain this understanding without emending the text.

In the discussion of the stones of fire, some scholars have drawn attention to the phrase *abn brq*, which appears in reference to the palace of Baal in the Ugaritic literature,[250] and the phrase *abn brq* has

---

[245]Gilgamesh IX; text adapted from Parpola, *The Standard Babylonian Epic of Gilgamesh*, 102, lines 175–78; translation according to George, *The Epic of Gilgamesh*, 75, lines 173–76.

[246]See my discussion of fruit-bearing trees in the divine abode, above in chapter 2.

[247]R. Dussaud, "Les Phéniciens au Négeb et en Arabie d'après un text de Ras Shamra," *RHR* 108 (1933): 40. For additional references see Pope, *El in the Ugaritic Texts*, 99.

[248]J. Steinmann, *Le prophète Ézékiel et les débuts de l'exil* (Paris: Editions du Cerf, 1953), 146–48. Steinman compares this to the flame of the sword in Gen 3:24. Cf. R. S. Hendel, "'The Flame of the Whirling Sword': A Note on Genesis 3:24," *JBL* 104 (1985): 671–74; P. D. Miller, "Fire in the Mythology of Canaan and Israel," *CBQ* 27 (1965) 256–61.

[249] H. G. May, "The King in the Garden of Eden," 166–76.

[250]*CTU* 1.3.III.26.

been related to the 'stones of lightning' which appear in an Akkadian prayer to the storm-god.[251] Cassuto made this association and found the reference to be related to the burning coals of fire (*gaḥălê ʾēš*) in Ezek 1:13 from which lightning came forth.[252] If Cassuto is correct, I find it difficult not to explain the *ʾabnê ʾēš* as deities in the divine council.[253]

The context of the *abn brq* in the Ugaritic Baʿal cycle is interesting:

> *dm . rgm / ʾiṯ . ly . w . ʾargmk*
> *hwt . w . ʾaṯnyk .*
> *rgm / ʿṣ . w . lḫšt . ʾabn /*
> *tʾant . šmm . ʿm . ʾarṣ /*
> *thmt . ʿmn . kbkbm /*
> *abn . brq . dl . tdʿ . šmm /*
> *rgm . ltdʿ . nšm*[254]
>
> For a word I have that I must speak to you,
> A message that I must repeat;
> A word of tree(s) and a whisper of stone(s),
> conversation of heaven with earth,
> of sea with stars,
> Stones of lightning unknown to the heavens,
> thunder[255] which people do not know;
> and the multitude of the earth does not understand.

Here the *abn brq* appear within the context of heaven, earth, sea, and stars conversing — a point to which I shall return shortly. There is some question as to how the phrase should be read. Prior to Cassuto's

---

[251]L. W. King, *Babylonian Magic and Sorcery*, (London: London, Luzac, 1896), 78, line 2 (No. 21, line 17). The phrase is *zunnu u abnē*^meš^ *birqu iš(ātu)* taken by King as 'rain and hail, lightning, fire'. The relationship between *abnē*^meš^ and *birqu* is not immediately clear.

[252]U. Cassuto, cited by Yaron, "The Dirge over the King of Tyre"; Cassuto, *The Goddess Anath: Canaanite Epics of the Patriarchal Age* (trans. Israel Abrahams; Jerusalem: Magnes, 1971), 90–91.

[253]This is noteworthy particularly in view of the description of the divine council meeting in 1 Kgs 22:19, which includes *kol-ṣəbāʾ haššāmayim* 'all the host of heaven'.

[254]*CTU* 1.3.III.20–27.

[255]Accepting Fensham's translation (cf. Akkadian *rigmu*) following a suggestion of W. F. Albright. See Fensham, "Thunder-Stones in Ugaritic," *JNES* 18 (1959) 274. Cassuto translates *rgm* 'matter' (*The Goddess Anat*, 91).

interpretation of *abn brq*, Virolleaud's original reading of the text took *abn* as deriving from the root *bnw*, 'to build', a suggestion which others followed, including Herdner, Gaster, and Driver.[256] Gordon construed a verb from the root *byn* 'to understand', which Gibson later adopted in his update of Driver.[257] F. C. Fensham arrived at an understanding like that of Cassuto, by taking the phase *abn brq* to mean 'thunder-stones', a reference to flint.[258] A consensus seems to have emerged around reading *abn* in *CTU* 1.3.III.26 as 'stone' as opposed to a verbal form meaning 'to build' or 'to understand'.[259] The references to heaven, earth, sea, and stars signify no less than divine manifestations, and it is in this sense that the phrase *abn brq* is perhaps best understood.[260]

I believe the best understanding of the *ʾabnê ʾēš* of Ezek 28:14 may lie in a combination of celestial and temple imagery. H. Jahnow and R. Wilson, for different reasons, interpret the stones of fire as signifying the coals of fire (*gaḥălê ʾēš*) in the temple. H. Jahnow suggested: "die feurigen Steine, in deren Kreise der Tempelwächter umherwandelt, sind wohl eine Art Waberlohe, die das Heiligtum auch noch als Schutz umgibt."[261] R. Wilson suggests that the *ʾabnê ʾēš* are the coals from the altar (*gaḥălê ʾēš*), and as such represent the

---

[256]A. Herdner, "Remarques sur la 'déesse ʿAnat'," *Revue des études sémitiques-Babyloniaca*, fasc. I (1942–45): 39; T. H. Gaster, *Thespis* (New York: Schuman, 1950), 215; G. R. Driver, *Canaanite Myths and Legends* (Edinburgh: T. & T. Clark, 1956) 87.

[257]C. H. Gordon, *Ugaritic Literature* (Rome: Pontifical Biblical Institute, 1949), 19, and *Ugaritic Manual* (Rome: Pontifical Biblical Institute, 1955), 246; J. C. L. Gibson, *Canaanite Myths and Legends* (Edinburgh: T. & T. Clark, 1978).

[258]According to Fensham ("Thunder-Stones in Ugaritic," 273), "The flint artifacts of prehistoric man make people think that they were produced by the gods and especially by the weather- or thunder-god."

[259]Pope, *El in the Ugaritic Texts*, 99; J. Oberman, *Ugaritic Mythology* (New Haven: Yale University Press, 1948), 26; H. L. Ginsberg translates 'a thunderbolt', *ANET*, 136.

[260]See Cross, "The Council of Yahweh in Second Isaiah," *JNES* 12 (1953): 275 n. 3 and discussion; and "'The Olden Gods' in Ancient Near Eastern Creation Myths" in *Magnalia Dei*, 329–38.

[261]The fiery stones, in whose midst the temple guard walked about, are probably a kind of blazing smoke, which surrounds the sanctuary even as protection. H. Jahnow, *Das Hebräische Leichenlied* (BZAW 36; Giessen: Alfred Töpelmann, 1923).

"glowing fire of Yahweh's presence."[262]

The apparent close connection of the two statements *ʾet-kərûb mimšaḥ hassōkēk* and *bətôk ʾabnê ʾēš hithallaktā* suggests that the stones of fire are none other than the coals of fire described in 10:2–4:

[2]בֹּא אֶל־בֵּינוֹת לַגַּלְגַּל אֶל־תַּחַת לַכְּרוּב וּמַלֵּא חָפְנֶיךָ גַחֲלֵי־אֵשׁ מִבֵּינוֹת לַכְּרֻבִים וּזְרֹק עַל־הָעִיר וַיָּבֹא לְעֵינָי׃ [3]וְהַכְּרֻבִים עֹמְדִים מִימִין לַבַּיִת בְּבֹאוֹ הָאִישׁ וְהֶעָנָן מָלֵא אֶת־הֶחָצֵר הַפְּנִימִית׃ [4]וַיָּרָם כְּבוֹד־יְהוָה מֵעַל הַכְּרוּב עַל מִפְתַּן הַבָּיִת וַיִּמָּלֵא הַבַּיִת אֶת־הֶעָנָן וְהֶחָצֵר מָלְאָה אֶת־נֹגַהּ כְּבוֹד יְהוָה׃

[2]"Go in among the wheels underneath the cherub (*ʾel-bênôt laggalgal ʾel-taḥat lakkərūb*, lit. 'to the midst of the wheels to the underside of the cherub'), fill your hands with ***burning coals from between the cherubim (gaḥălê ʾēš mibbênôt lakkərūbîm)***, and scatter them over the city." And he went in before my eyes. [3]Now the cherubim were standing on the south side of the house, when the man went in; and a cloud filled the inner court. [4]And the glory of Yahweh went up from the cherub to the threshold of the house; and the house was filled with the cloud, and the court was full of the brightness of the glory of Yahweh.[263]

Here, the coals of fire (*gaḥălê ʾēš*) are described as being in the midst of the cherubim. The description is made with regard to the temple, and the language recalls that of the storm theophany. This interpretation of the *ʾabnê ʾēš* is also suggested by the description of the *gaḥălê ʾēš* of Ezek 1:13:

וּדְמוּת הַחַיּוֹת מַרְאֵיהֶם כְּגַחֲלֵי־אֵשׁ בֹּעֲרוֹת כְּמַרְאֵה הַלַּפִּדִים הִיא מִתְהַלֶּכֶת בֵּין הַחַיּוֹת וְנֹגַהּ לָאֵשׁ וּמִן־הָאֵשׁ יוֹצֵא בָרָק׃

As for the image of the living creatures: their appearance was like burning coals of fire, like the appearance of torches, moving about amid the living creatures. And the fire was bright, and from the fire lightning was going forth.[264]

---

[262]R. R. Wilson, "The Death of the King of Tyre: The Editorial History of Ezekiel 28," ***Love and Death in the Ancient Near East: Essays in Honor of Marvin H. Pope*** (ed. J. H. Marks and R. M. Good; Guilford, Conn.: Four Quarters, 1987), 211–18. Wilson cites inter alia Ezek 10:2; Lev 16:12.

[263]Note that several times the wheels are said to be beside the wings (3:13; 10:19).

[264]G και εν μέσω...ὅρασις and OL suggest the reading ובתוך...מראה for ודמות מראיהם. The ה of הלפדים is most likely due to dittography and should be omitted.

Here, Ezekiel likens the *appearance* (*marʾeh*) of the living creatures to burning coals of fire. He describes the *form* (*dəmût*) of the living creatures as resembling what looks like *torches moving about* (*mithalleket*) *their midst*. In Ezek 28:14, the primal human is with the cherub and 'moves about' (*hithallāktā*) *bətôk ʾabnê ʾēš* 'amid stones of fire'. In 1:13, the 'appearance of torches' (the apparent subject of *mithalleket*) 'moves about' *bên haḥayyôt* (amid the creatures).[265] The four creatures are none other than cherubim, the composite beings that surround the divine throne. Although they are not so named, their identity is clear from the description in chapter 10, noted above (see also Exod 25:19–22; 1 Kgs 6:23–28; cf. 2 Kgs 19:15). The use of *mithalleket* in the description is interesting. The text has signs of corruption, but the word expresses divine activity in several contexts (see Gen 3:8; Job 1:7=2:2; 22:14; Zech 1:10,11; 6:7). The cherub and the primal human are associated with the stones of fire once again in v. 16, where the cherub 'drives' the primal human from the stones of fire.[266]

There appears to be a clear interplay among cherubim, altar coals, and celestial phenomena. The symbolism underlying this is not difficult to discern. The heavenly host represents a concrete manifestation that reflects the glory of Yahweh (e.g., Ps 19:2–5; 148:1–6 [v. 6 cf. Deut 32:8]); as members of the divine court, the celestial 'armies' (*ṣəbāʾôt*) attend and attest to his presence (see Job 38:7).[267] In much the same way, the coals which illumine the mercy seat (Lev 16:12–13) are a concrete sign of Yahweh's presence and could well be considered another manifestation of the divine courtly deities. The smoke and fire

---

[265] The grammatical problem of agreement of gender, reconciling *hîʾ mithalleket* (f) with either *marʾeh* or *lappîd* (both m.), cannot be corrected by taking *dəmût* (f.) as the subject of *hîʾ mithalleket* without rendering the passage nonsensical. BDB seems to take *gaḥălê* 'coals' as feminine, citing it under the extant feminine form *gaḥelet* (160). This too is not without problems, but presents another possible antecedent of *hîʾ mithalleket* that supports the point made here.

[266] Further indication that the *ʾabnê ʾēš* here may refer to deities may be found in the observation that the banishment from the stones of fire in v. 16 recalls the fall of "Day Star" from among the other stars in Isa 14:12–15.

[267] The stars generally refer to the celestial armies of Yahweh, who is portrayed as warrior. Job 38:7 demonstrates that the reference to stars is not confined to the warrior/armies image. See also 1 Kgs 22:19–20.

which proceed from them invoke the images of clouds and lightning which characterize the storm theophany. This is evident in Psalm 18 = 2 Samuel 22 (vv. 9–14):

9 עָלָה עָשָׁן ׀ בְּאַפּוֹ
וְאֵשׁ־מִפִּיו תֹּאכֵל
גֶּחָלִים בָּעֲרוּ מִמֶּנּוּ׃
10 וַיֵּט שָׁמַיִם וַיֵּרַד
וַעֲרָפֶל תַּחַת רַגְלָיו׃
11 וַיִּרְכַּב עַל־כְּרוּב וַיָּעֹף
וַיֵּדֶא עַל־כַּנְפֵי־רוּחַ׃
12 יָשֶׁת חֹשֶׁךְ ׀ סִתְרוֹ סְבִיבוֹתָיו
סֻכָּתוֹ חֶשְׁכַת־מַיִם עָבֵי שְׁחָקִים׃
13 מִנֹּגַהּ נֶגְדּוֹ
עָבָיו עָבְרוּ בָּרָד וְגַחֲלֵי־אֵשׁ׃
14 וַיַּרְעֵם בַּשָּׁמַיִם ׀ יְהוָה
וְעֶלְיוֹן יִתֵּן קֹלוֹ בָּרָד
וְגַחֲלֵי־אֵשׁ׃

9 Smoke went up from his nostrils
and devouring fire from his mouth;
glowing coals (*geḥālîm*) flamed forth from him.
10 He bowed the heavens, and came down;
thick darkness was under his feet.
11 *He rode on a cherub and flew;*[268]
he came swiftly upon the wings of the wind.
12 He made darkness his covering around him,
his canopy thick clouds dark with water.
13 Out of the brightness before him
there broke through his clouds hailstones and coals of fire (*gaḥălê ʾēš*).
14 Yahweh also thundered in the heavens,
and the Most High *uttered his voice,*
hailstones and coals of fire (*gaḥălê ʾēš*).

Similar imagery is found in Ps 104:4: '[You] who make the clouds your chariot, who ride on the wings of the wind, who make the winds your messengers; fire and flame your ministers'.[269] Here, the imagery

---

[268] Italics in this passage indicate emphasis, not an uncertain translation.

[269] We recall that in Ezek 10:2 (cited above), the coals are 'scattered over the city'

is personified in the role of the *malʾāk* 'messenger' and the *məšārēt* 'minister'.

It is also significant for our purposes to note that God "speaks" directly with humanity through the storm imagery. As we have seen, Ezekiel's vision of the chariot is couched in the language of the storm theophany (see 1:4,5), and it is relatively clear that the storm theophany is emulated in the blazing fire and smoke of the sanctuary. Just as God spoke to Moses from the midst of the burning bush on Sinai (cf. Elijah on Horeb, the mountain of God, 1 Kgs 19:8–12), so did he speak from the midst of the cherubim (Exod 25:22; Num 7:89 *mibbên šnê hakkərûbîm*; cf. Ps. 29:7). This imagery of the storm and of smoke and fire, and of speaking through these media, is clearly combined in Isa 30:27:

הִנֵּה שֵׁם־יְהוָה בָּא מִמֶּרְחָק
בֹּעֵר אַפּוֹ וְכֹבֶד מַשָּׂאָה
שְׂפָתָיו מָלְאוּ זַעַם
וּלְשׁוֹנוֹ כְּאֵשׁ אֹכָלֶת׃

Behold the name of Yahweh comes from afar,
burning with his anger, and in thick rising smoke;
his lips are full of indignation ,
and his tongue is like a devouring fire.[270]

The same is true in Isa 30:30: 'and Yahweh will cause his majestic voice to be heard and the descending blow of his arm to be seen, in furious anger and a flame of devouring fire, with a cloud burst and a tempest of hailstones' (cf. Exod 19:18). The connection with the sanctuary is explicit in Isa 6:5. In Isaiah's temple vision, when Yahweh speaks, the sanctuary fills with smoke: 'And the foundations of the thresholds shook at the voice of him who called, and the house was filled with smoke'. The imagery here is of a roaring fire, which, as it is stoked, sends up billows of smoke.[271]

— an image of divine judgment.

[270]Cf. Ps. 29:7.

[271]Compare, however, Haran, "an altar, upon which sacrifices were offered up (in contrast to the incense altar, which is a separate matter) was found only in the open." Note that Haran does not comment on the significance of the fire and coals under the cherubim in 10:7 (*Temples and Temple Service*, 15).

Indeed, the glory of Yahweh manifests itself in the fire of the sanctuary; and the beings most closely associated with this manifestation are the cherubim and seraphim, both of which are intimately associated with fire. The call of the seraphim in Isaiah 6 — 'the whole earth is full of his glory' — refers to the filling of the sanctuary with flaming smoke. In the case of both the temple fixtures and the heavens, mundane phenomena have sacred referents. The sensitive Israelite hearer would certainly have perceived both the reference to that which the *ʾabnê ʾēš* signified in a concrete sense (coals) and to what they implied beyond that (deities). Such is the nature of religious discourse.

Pope was, I believe, only partially correct in stating that despite the obscurity that surrounds the meaning of the *ʾabnê ʾēš*, "it is clear that it refers to the splendiferous surroundings" described by the narrator.[272] More importantly, the stones of fire are indicative of divine presence, and the language draws attention to the exalted nature and position of the primal figure, specifically with respect to his relation to deity. Difficulties of the text notwithstanding, it is clear that in the first section of this oracle (vv. 11–14) the figure is described in exceedingly lofty terms. He is associated with temple imagery and appears to be closely associated with Yahweh himself.

The "Fallen" State of the Primal Human (15–18)

Transition to new condition (15)

As I stated at the outset of our discussion of 28:11–19, v. 15 stands as a transitional verse between the two major sections of the oracle, and describes the movement of the figure from one state to the other. Regardless of how one chooses to analyze the meter of this verse, it is clear that it marks a transition between the flow and content of what precedes it and what follows. In its present state, the section comprising vv. 12b–14 has a prose character.[273]

---

[272]Pope, *El in the Ugaritic Texts*, 99.

[273]The very different reconstructions of Zimmerli (*Ezekiel* 2.81–89) and Cooke (*Ezekiel*, 315–20) demonstrate the extent of the emendations required to offer a reasonable analysis of meter. Both analyses make sense, but only at the expense of the

The prose character of the first section shifts to a more easily recognized parallelism *after* v. 15, which, perhaps, may be considered a tricolon (as the editors of NRSV have done), but nonetheless is very prosaic:

תָּמִים אַתָּה בִּדְרָכֶיךָ
מִיּוֹם הִבָּרְאָךְ
עַד־נִמְצָא עַוְלָתָה בָּךְ׃

You were blameless in your ways
from the day you were created,
until iniquity was found in you.

In this verse, the previous condition is stated, and in the same breath conflict is introduced. What is clear here is that the prophet is still laying out the essential structure of the myth. Within the context of the passage, the verse reads as the last in a string of uninterrupted mythological statements. As it stands, v. 15 represents a transition between positive and negative statements about the protagonist. Moreover, it represents a transition between statements concerning the natural endowments and situation of the figure (vv. 12b–14) and statements concerning the figure's *conduct* (vv. 16–18). The notion of conduct is introduced for the first time in the oracle in v. 15, which begins with the positive statement, 'You were blameless in your ways', and closes with the negative statement, 'until iniquity was found in you', introducing further negative statements about conduct and the resulting consequences. It also marks a point at which more elements of the tenor of the metaphor are introduced. For the next three verses, the third colon (or perhaps better "statement") of v. 15 ('iniquity was found in you') is explicated in a stylistically repetitious manner. There, the prophet reveals the import of his analogy. I will address the third section below.

The statement *tāmîm ʾattâ bidrākeykā* 'you were blameless in your ways' is one of the highest tributes paid to an individual in the Hebrew Bible. There are those who are spoken of in rather impersonal

---

text itself. In this case, I find metrical analysis too circular to be of value — nothing may be learned of the so-called *qînâ* meter, for the text has been straightjacketed into it.

terms who are so described (Ps 101:2; 119:1; Prov 11:20); it is not often, however, that the term is linked to a specific individual. In Gen 6:9 it is applied to Noah, who is himself presented as a primal human figure, and it is the precondition and reason for the salvation of humanity from the flood.[274] In Gen 17:1 it underlies the covenant between Abram and El Shaddai: the covenant is predicated on the condition that Abram be 'blameless' *tāmîm*. One thinks also of the description of the righteous sufferer Job in the prologue of that book (1:1).[275]

Statement of judgment (16–18)

As stated above, the third colon of v. 15 is explicated in a stylistically repetitious manner in vv. 16–18. The sin is given in the first "half" of each statement and Yahweh's corrective response is given in the second half. This construction may be readily seen by placing the verses side by side:

v. 16	v. 17	v. 18
In the abundance of your trade You were filled with violence;	Your heart was proud because of your beauty;	By the multitude of your iniquities, in the unrighteousness of your trade,
And *you sinned*	*you corrupted* your wisdom for the sake of your splendor	*You profaned* your sanctuaries
So *I cast you* as a profane thing from the mountain of God	*I cast you* to the ground;	*I brought forth fire* from the midst of you; it consumed you,
And *the guardian cherub drove you* out from the midst of the stones of fire.	*I set you* before kings so they could look upon you.	and *I turned you to ashes* upon the earth in the eyes of all who see you.

[274]See the discussion of Noah as primal human below in chapter 7.

[275]This verse employs the verb of the same root, rather than the adjective.

In this section tenor, i.e., the king of Tyre, and vehicle, i.e., the primal human figure, are woven together, and it is difficult to detect one from the other. Statements specific to Tyre, which are not a part of the mythic tradition but instead only explicate the use of the tradition, are not easily discerned. Verse 16b is an unequivocal reference to the mythic complex. Other statements in vv. 16–18 certainly interact with the mythic complex, but the extent to which any other of these presents a *direct* allusion is debatable.

Trade and violence (16a)

בְּרֹב רְכֻלָּתְךָ מָלוּ תוֹכְךָ חָמָס
וַתֶּחֱטָא

> In the abundance of your trade you were filled with violence,
> and you sinned.

In the structure of vv. 16–18, trade (vv. 16,18) is juxtaposed with beauty (v. 17). This is no surprise, given the fact that Tyre's beauty was explained with reference to trade in the oracle of chapter 27 (especially vv. 2–10). If the beauty of Tyre is related to trade, as I believe is quite clear, then it becomes clear that in the structure of 16–18, the concept of trade is modified by violence (*ḥāmās* v. 16), pride (*gābah libbəkā* v. 17), and iniquities and unrighteousness (*ᶜăwôneykā*, *ᶜewel* v. 18).

Tyre was, of course, a city known for its economic strength developed through international trade. This was duly noted by the prophet throughout chapter 27. The language of chapter 27 suggests that the lament of 28:11–19 is directed to the city of Tyre. More importantly, this understanding implies that the mythological analogy is not drawn concerning the king as an individual exclusively, but posits him as a representative of the city. The phrase 'you sinned' applies directly to Tyre, and, most likely also belongs to the structure of the myth. The references to trade and violence explain the application of the analogy to Tyre.[276]

---

[276]The reference to trade and violence is considered by most critical scholars a later explanatory gloss.

Was trade included in the version of the primal human story used by Ezekiel? It goes virtually without saying that the primal human was *not* involved in trade in the tradition to which Ezekiel appeals; at least we have no corroborating evidence to suggest as much. It is possible that there is something about trade that parallels something in the primal human tradition, but it is also possible that there is no such parallel, and that it was simply drawn from other messages directed toward Tyre. As such, the trade reference constitutes little more than a rehearsal of the standard complaint against Tyre, perhaps stated to clarify the meaning of the analogy.

The notion of being full of 'violence' is tied to trade in Ezek 28:16. Throughout the Hebrew Bible, the word that appears most often in parallel to *ḥāmās* is *šādād*, which refers to the oppression of the poor. Ezekiel's understanding of the term is expressed in 45:9: 'Thus says Yahweh God: Enough, O princes of Israel! Put away violence (*ḥāmās*) and oppression (*šōd*), and execute justice and righteousness; cease your evictions of my people, says Yahweh God'. The words used to translate *ḥāmās* in the Septuagint suggest the primary sphere of interest is disobedience to the law and injustice.

'Profaned' from the mountain/removed from the stones of fire (16b)

וָאֲחַלֶּלְךָ מֵהַר אֱלֹהִים
וָאַבֶּדְךָ כְּרוּב הַסֹּכֵךְ מִתּוֹךְ אַבְנֵי־אֵשׁ׃

> So I cast you as a profane thing from the mountain of God,
> and the guardian cherub drove you out from the midst of the stones of fire.

One of the more common analogues to this verse is found in Isaiah 14. In it, the protagonist who aspires to a throne on the mount of assembly is, instead, brought down to Sheol, to the depths of the Pit (14:13–15). To be cast from the mountain is to be taken from the place where the divine could be encountered. The phrase is literally rendered 'I profaned you *from* the mountain'. Scholars have been largely silent on the use of this word here, despite its apparent strangeness in translation. It is no less than a statement of excommunication.

The verb *ḥll* is closely associated with the cult.[277] The root *ḥll* is regularly used as an opposing term to the root *qdš* in the distinction of sacred and profane, and they are found as parallels to *ṭmʾ* and *ṭhr*.[278] This verse is one of the few times in which it is said that a person is profaned by the deity. In Isaiah 43, the cultic sins of Israel are recounted (vv. 22–27). The consequence of these sins is expressed in v. 28: 'Therefore I profaned the princes of the holy place' (*waʾăḥallēl śārê qōdeš*). This is certainly a dismissal of cultic officials from the sacred space; they are disqualified from service by virtue of their status as unclean. The fact that Yahweh has done this deliberately — it is an "act" that he performs, and they are, as it were, "declared" unclean — lends to the ritual flavor of the situation. The use of the term *qōdeš* 'holy place' is particularly interesting in view of the 'mountain' of Ezek 28:16. Of the occasions in which *qōdeš* is used with reference to a sacred place, Ps 20:3 uses it to signify (or, associates it with) none other than the mountain, Zion: 'May he send you help from the holy place (*qōdeš*), and give you support from Zion'.[279] The choice of word *ḥll* by the prophet, then, is not difficult to understand. It is deliberate and is in keeping with the imagery presented throughout the dirge.

The verb used to describe the apparent removal of the first man from the stones of fire is *ʾbd* 'to perish, be destroyed'. The Piel of this root is used elsewhere in Ezekiel to refer to the destruction of high places ('I will destroy your high places', 6:3) and the destruction of life ('Her princes in the midst of her are like wolves...destroying lives to get dishonest gain', 22:27). The verb appears in reference to the fall of Tyre in 26:17: 'How you have perished ... from the seas' (*ʾêk ʾābadt ... miyyammîm*). Although the verb in 26:17 is in the Qal stem, it resembles the occurrence in 28:16 insofar as both use the preposition *min-* 'from'. The significance of the use of the verb then is that it is tied to the notion of punishment, or the consequences of entering into a

---

[277]Particularly when it appears in the Piel. The original meaning is thought to be 'to loosen, untie', which appears in the Akkadian cognate *elēlu* 'to be free, set free' (*AHw*, 1.197). The Akkadian term also means 'to be pure, purify' with reference to the cult.

[278]For references and discussion, see W. Dommershausen, "*ḥll*," in *TDOT* 4.416.

[279]NRSV appropriately translates *qōdeš* 'sanctuary'.

conflict with Yahweh. The root *ʾbd* in the Qal can also mean 'to wander, go astray' (Deut 26:5; 1 Sam 9:3); it is thus also possible to construe its use in Ezek 28:16 as a factitive expression of the intransitive sense of the Qal — 'made you wander'. This recalls the situation of Cain in Gen 4:14, according to which he is driven (*grš*) from the earth, hidden (*str*) from the face of Yahweh, and destined to be a fugitive and a wanderer (*nāʿ wanād*).[280] Thus, there is an inherent ambiguity that heightens the effect of the expression in Ezek 28:16. The removal from the stones of fire is the banishing of the primal human from the realm of the deities. He would no longer walk among them, and as the text tells us, death would be his lot.

Hubris and corrupted wisdom (17a)

גָּבַהּ לִבְּךָ בְּיָפְיֶךָ
שִׁחַתָּ חָכְמָתְךָ עַל־יִפְעָתֶךָ

Your heart was proud because of your beauty,
You corrupted your wisdom for the sake of your splendor.

There is an obvious literary connection between Ezek 28:3 and 28:17: the phrase 'your heart was proud' occurs in both. Hubris seems to refer in these verses to the misappropriation of wisdom (the divine prerogative). We have seen in the description of the primal human in this second oracle of Ezekiel 28 that wisdom is ascribed to him. In the first oracle, this wisdom is noted to be unequaled (vv. 3–5). What is at issue is the fact that this privileged position and the accouterments of office have been parlayed into personal wealth, at the expense of those who should rather benefit from such things to which the mediating figure has access. It is an abuse of the office of the mediator.

Cast 'upon the ground' before kings (17b)

עַל־אֶרֶץ הִשְׁלַכְתִּיךָ
לִפְנֵי מְלָכִים נְתַתִּיךָ לְרַאֲוָה בָךְ

---

280 On the interpretation of Cain as a primal human figure, see my comments in chapter 7.

I cast you to the ground;
I set you before kings to feast their eyes on you.

This is imagery that represents death and not a simple pushing down. We find a similar expression in Ps 17:11, spoken of the enemies of David: 'They set their eyes to extend me to the earth' (*ʿênêhem yāśîtû linṭôt bāʾāreṣ*). The two clauses also parallel those in v. 18, according to which the protagonist is made 'ashes'. We will return to the significance of this below in our discussion of v. 18.

The casting down upon the earth here is done in the sight of kings. The reason is, in part, to make the protagonist an example. It is more than likely that there is significance in this reference with respect to the primal human. We will discuss the implications further below.

Unrighteous trade and profaning the sanctuary (18a)

מֵרֹב עֲוֺנֶיךָ בְּעֶוֶל רְכֻלָּתְךָ
חִלַּלְתָּ מִקְדָּשֶׁיךָ

By the multitude of your iniquities in the unrighteousness of your trade
you profaned your sanctuaries.

The charge of profaning sanctuaries is curious. What would a Yahwistic prophet care about the ritual correctness of a foreign ruler with respect to his foreign god? One can also ask this question of the earlier statement in v. 15, 'You were blameless in your ways'. The statement, 'You profaned your sanctuaries', can also be understood, however, as belonging to the essence of the myth. If "sanctuary" can be understood as homologous with the concept of garden, then the mythic reference can involve some sort of offense in the garden.[281]

This is not the only place where Ezekiel expresses Yahweh's concern over the profaning of sanctuaries. In 5:11, the charge is leveled against Israel: 'Because you have defiled my sanctuary (*ʾet-miqdāšî ṭimmēʾt*) with all your abominations, therefore I will cut you down...'.

[281] We can similarly suggest concerning Genesis that disobedience in the garden is tantamount to profaning the sanctuary. R. Wilson ("The Death of the King of Tyre," 211–18) has interpreted Ezek 28:11–19 as an "oblique" oracle directed against the priests of Israel, based, in part upon this verse. Wilson's is an intriguing suggestion, but the fact that it is addressed to Tyre remains a difficulty for his analysis.

Likewise, in 23:38–39 he writes: 'They have defiled my sanctuary (*ʾet-miqdāšî ṭimmәʾû*) on the same day and profaned my sabbaths (*wәʾet-šabbәtôtay ḥillēlû*). For when they had slaughtered their children in sacrifice to their idols, on the same day they came into my sanctuary (*miqdāšî*) to profane it (*lәḥallәlô*)'.[282]

Turned to ashes (18b)

וָאוֹצִא־אֵשׁ מִתּוֹכְךָ הִיא אֲכָלַתְךָ
וָאֶתֶּנְךָ לְאֵפֶר עַל־הָאָרֶץ לְעֵינֵי כָּל־רֹאֶיךָ

So I sent forth fire from the midst of you; it consumed you
and I turned you to ashes upon the earth in the eyes of all who see you.

There is more than likely a double referent in the phrase 'turned to ashes'. The more prominent reference is to the city of Tyre in its apparent reduction to rubble through conflagration. The other regards the king as primal human. This mythical dimension is clearly evident in the many references in the Hebrew Bible, and indeed throughout the ancient Near East, to dust and the creation and demise of humanity.

The burning of a city was a common phenomenon in antiquity. Poets regularly drew upon this imagery to express war and judgment.[283] The language used in v. 18 is typical of prophetic oracles found elsewhere in the Hebrew Bible. Num 21:28, for example, records a saying of ballad singers that 'fire went forth (*yāṣәʾâ*) from Heshbon, flame from the city of Sihon. It devoured Ar of Moab, the lords of the heights of the Arnon'. Thus, there is nothing difficult or particularly unusual about the phrase here in Ezekiel, 'I sent forth fire from your midst' (*wāʾôṣîʾ-ʾēš mittôkәkā*). One difference between Num 21:28 and Ezek 28:18, however, is the reference in Ezek 28:18 to becoming ashes: Ezek 28:18 is framed within the context of the primordial human, and the allusion to ashes invokes the idea of the being's humanity.

It may be that Ezekiel is making an allusion to the thirteen-year siege of Tyre by Nebuchadrezzar, which is laid out clearly in chapter

---

[282]Cf. Ezek 24:20; 25:3.

[283]For a discussion of the use of fire imagery in the Hebrew Bible and related literatures see V. Hamp, "*ʾēsh*," in *TDOT* 1.418–28.

26.[284] The siege lasted from ca. 586–573 B.C.E. The description of the siege in chapter 26, though detailed, makes no mention of a conflagration, and the city was certainly not destroyed. It may be that the writer had another specific event in mind, but there is no evidence to suggest what that might have been. Amos records a similar situation: 'I will send fire upon the wall of Tyre, and it shall devour (*wə-ʾākəlâ*) its strongholds' (Amos 1:10). Whatever the reference, the event seemed to have no permanent effect; Tyre continued to prosper. The fact that it is past tense, unlike the similar statements of Amos regarding Damascus and Gaza (Amos 1:4,7), suggests that it is anchored in historical reality.

It is possible that the statement is prophetic rhetoric, completely void of any historical referent, and that the best way to understand the reference to fire is as a common prophetic metaphor for judgment. The imagery would be a reference to destruction, but not necessarily with reference to any specific structure. The destruction of strongholds is not an end in itself — rather, it is a metaphor for rendering a city defenseless, and constitutes an end to life as it was once known. There are abundant examples throughout the Hebrew Bible of fire being used to express divine judgment. Ezekiel employs fire to express judgment against Judah (19:14; 21:3); the Ammonites (21:37); the house of Israel (22:21); Jerusalem (23:25; cf. 24:10); Egypt (30:8,14,16); and Gog (38:22). In three of these passages, the prophet makes reference to the fire consuming (*ʾkl*) the subject (21:3; 21:37; 23:25). These verses and other similar references demonstrate that the threat of destruction by fire was a stock image of the prophetic repertoire. The basis of the imagery was no doubt fixed in the violent realities of war; however, it seems that the point is judgment and destruction, and not necessarily the literal fire itself.[285]

The reference to ashes may be applied to the city, but the context of the passage suggests that it may also apply to the king. The myth used as the basis for the lament regards an individual, and the lament

---

[284]See Josephus, *Ant.* 1.228; *Contra Apion* 1.156.

[285]The figurative nature of fire for judgment is also made clear in 38:22, in which fire and sulfur rain down on Gog (see also Gen 19:24).

is directed to the king specifically, as figurehead and representative of the nation. All of the other statements in vv. 16–18 apply easily to an individual. When taken as such, the reference to ashes takes on new significance, for it is of the very language used to express the mortality of humanity in Gen 2:7; 3:19.

The fact that the term *ʾēper* is used, as opposed to *ʿāpār* of the Genesis account, is of little consequence. The two terms are interchangeable as far as they regard the composition of the human body. Job likened his situation to death, saying that he had been cast into the clay (*ḥōmer*) and had 'become like dust (*ʿāpār*) and ashes (*ʾēper*)', all three words vividly expressing his physical constitution (Job 30:19).[286] In Gen 18:27, Abraham expresses a state of humility before God with the words, 'I have set out to speak to Yahweh, though I am but dust (*ʿāpār*) and ashes (*ʾēper*)'. It is clear that no distinction is to be made here between the two terms *ʾēper* and *ʿāpār*. But this statement in Gen 18:27 is interesting for another reason: in it, Abraham's expression of humility (underscored by the use of the Hiphil verb *hôʾaltî* 'I have set out/I have taken it upon myself', which is often used as an expression of politeness or modesty) is accomplished by contrasting God and humanity. The physical composition of humanity clearly captures an essential difference between humanity and the divine.[287]

The reference to ashes in Ezek 28:18 seems quite clearly to be an intentional addition to the imagery of fire as judgment. Elsewhere, in Ezekiel and throughout the Hebrew Bible, the imagery of fire as judgment does not refer to the end result as 'ashes' (cf. Am 1:4,7,10,12,14; 2:2,5; Isa 10:17; 30:30; Jer 49:27; 50:32). The use of this phrase is also a clever way of introducing the concept of the physical make-up of the first human.

---

[286]Note also the close connection of clay and ashes in Job 13:12. For another discussion of dust imagery, see D. R. Hillers, "Dust: Some Aspects of Old Testament Imagery," *Love and Death in the Ancient Near East*, 105–9.

[287]Mal 3:21, 'You will tread upon the wicked, for they will become ashes (*ʾēper*) under the soles of your feet', is another example of the term *ʾēper* used to signify the essence of humanity. The image of ashes under the feet of the righteous victors represents the death of the trampled wicked. Here again, the focus on the material highlights the delicate and ephemeral nature of human life.

The phrase, 'I turned you to ashes', may thus be understood as a reference to the essential humanity of the being. This conclusion is reinforced by the fate of the first man recorded in Gen 3:19: 'By the sweat of your brow you shall eat bread until you return to the ground (*ʾel-hāʾădāmâ*), for from it you were taken. You are dust (*ᶜāpār*) and to dust (*ʾel-ᶜāpār*) you shall return'. The fate assigned to the first man in Genesis was a direct result of his sin, as is also true in this section of Ezekiel. The fact that the ashes are qualified as being 'on the earth' (*ᶜal hāʾāreṣ*) seems to drive home the notion of the unity of humankind, ashes, and earth.[288] In Genesis and in Ezekiel, the primal human, a created being, errs with respect to wisdom and is consequently returned to that from which he was created.[289]

Conclusion (19)

The reference to awestruck onlookers in v. 19 (***kol-yôdəᶜeykā bāᶜammîm šāməmû ᶜāleyka ballāhôt hāyîtā*** 'All who know you among the peoples were awestruck over you. You became an object of terror') is a statement regarding the efficacy of the word of Yahweh and of his sovereignty.[290] It is an expression of the protagonist's being made an example. The final phrase of the oracle *wəʾênəkā ᶜad-ᶜôlām* 'and you ceased to exist forever', is interesting, but its mythological significance is difficult to assess. A clue may be found in the similar language used of Tyre in 26:21 (***ballāhôt ʾettənēk wəʾênēk ûtəbuqšî wəlōʾ timmāṣəʾî ᶜôd ləᶜôlām*** 'I will make you an object of terror and you will cease to exist; though you are sought, you shall never again be

---

[288]Cf. the punishment of Korah, Dathan, and Abiram, whom, according to Numbers 16, the ground swallowed up: 'They went down alive into Sheol...and they perished from the midst of the assembly' (*wayyōʾbədû mittôk haqqāhāl*; vv. 32–33). Of the remainder of the unholy priests, the text relates that 'fire came forth (*ʾēš yāṣəʾâ*) from Yahweh and consumed [them]' (v. 35; cf. Ezek 28:18 *wāʾôṣīʾ-ʾēš*). See my comments above on Korah, Dathan, and Abiram and the unholy priests (Num 16:35 and 26:10; cf. Lev 10:2; Num 11:3; 2 Kings 1:10–14).

[289]Although the text in Ezekiel does not state explicitly that the primal human was created from some aspect of the earth, the image is so widespread that we may cautiously suggest that it is assumed here.

[290]See W. Zimmerli, ***Ezekiel*** 2.94.

found') and 27:36 (*ballāhôt hāyît wəʾēnēk ʿad-ʿôlām* 'You became an object of terror and you ceased to exist forever'). 26:19–20 speaks of Tyre's being covered by waters and thrust into the pit, a description that is similar to the death of the primal human in the oracle of Ezek 28:1–10; the passage's mythological orientation appears in references to 'the people of old' (*ʿam ʿôlām*) and 'primeval ruins' (*ḥŏrābôt mēʿôlām*). It is difficult to determine the extent to which such mythological undertones were perceived in 28:19.

Summary of transgression and consequences

In the first half of 28:16–18 the text discusses the transgression of the primal human. We are told explicitly that he *sinned* (*teḥĕṭāʾ*) and this assertion is expounded upon. We learn that the sin is not trade, but *violence* (*ḥāmās*) in trade. The problem had also been referred to as *unrighteousness* (*ʿāwel*) in v. 15. The sin is, in essence, an abuse of his exalted situation. We learn that his heart became proud (*gābah libbəkā*) because of his beauty (*bəyopyekā*) and that he corrupted his wisdom (*šiḥattā ḥokmātəkā*), also for a reason having to do with beauty (*ʿal-yipʿātekā*). The text suggests — particularly when read alongside the previous oracle in 28:1–10 — that it was his wisdom that made possible the incredible success of the trade. The *result* of the sin was an incongruity whereby the place of his habitation, the holy mountain of God became 'profaned'.

The second half of vv. 16–18 discusses the consequences of the sin of the primal human: expulsion and death. He is 'profaned' from the divine habitation, and driven from among the divine inhabitants, the 'stones of fire'. His expulsion from the holy mountain was accompanied by a death sentence. He is cast to the earth (*ʿal-ʾereṣ hišlaktîkā*), an image of death, and is reduced to ashes (*ʾēper*). Both the earth and the ashes recall the physical make-up of the primal human and his return to that from which he was brought forth.

Significance of the Primal Human Theme

The primal human as a priestly or intermediary figure

It is quite evident that the tradition upon which Ezekiel bases his lamentation in 28:11–19 understands the primal human in priestly terms, or, perhaps better put, in "intermediary" terms. The imagery he employs is consonant with that of the sacral king, endowed for service as vice-regent of God and mediator between human and divine. This is evident first in the homology of the garden-and-mountain with the temple, which we discussed in chapter 2, and which is manifestly evident here. The coalescence of garden and temple is expressed elsewhere by Ezekiel in his vision of the temple in chapters 40–48.[291] Although he does not state the association explicitly in Ezekiel 28, other allusions are supplied which demonstrate it. In v. 13 the primal human possesses the stones of the high priestly pectoral, used in conjunction with obtaining divine knowledge. In v. 18 Ezekiel alleges that the king of Tyre 'profaned his sanctuaries'. The priestly orientation of the primal human imagery is reminiscent of the Mesopotamian Adapa who is a priest of Eridu.

There is no real difficulty over the fact that this dirge is composed for a king, for the imagery suggests a sacral king, that is, there is a clear conflation of king and priest in reference to the primal human. The perceived connection between the two in ancient Israel appears frequently in the Hebrew Bible. Melchizedek, for example, is known both as the king of Salem and a priest of El Elyon (Gen 14:18–20). In Ps 110:4, the psalmist proclaims of the king, 'You are a priest forever in the order of Melchizedek'. Saul performs several priestly actions including building an altar and using the Urim and Thummim (1 Sam 14:35, 41).[292] David too inquires of Yahweh by using the ephod (1 Sam 30:7); he offers burnt and peace offerings (2 Sam 6:17; Solomon also in 1 Kgs 3:3–4 et al.); he blesses the people in the name of Yahweh (2 Sam 6:18); and it is said in no uncertain terms that 'David's sons

---

[291]See my discussion "Eden: the Archetype of the Temple," above in chapter 2.

[292]Yaron ("The Dirge over the King of Tyre," 28–57) adds the prohibition on eating declared by Saul in 1 Sam 14:24.

were priests' (2 Sam 8:18). Amaziah says of the temple at Bethel, 'It is the *king's temple,* and it is the king's court' (Amos 7:13).

Thus the perceived connection between king and priest supplies the answer to the apparently mixed images in Ezekiel 28, and this connection can be explained, in part, on the basis of the ancient Near Eastern topos of the endowing of the king or hero with divine and sacral attributes at his creation.

Summary of Primal Human Topoi

The four themes or common threads I have identified which help to define the core of the primordial human tradition are, again, location, wisdom, creation, and conflict. These themes remain evident here in Ezek 28:11–19.

Creation and the primal human

Two images strongly invoke the setting of creation. In 28:13 *bəʿēden gan-ʾĕlōhîm hāyîtā* 'You were in Eden, the garden of God', seems best understood as a reference to the epoch of creation. In the same verse the phrase, *bəyôm hibbāraʾăkā kônānû* 'in the day you were created, they were established', confirms this, as does the phrase in v. 15, *miyyôm hibbārāk* 'from the day you were created'. The divinatory stones of the high priestly breastpiece and the timbrel and pipes that were 'established' on the day of his creation seem significant, in originating in the primordium. The phrase *bəyôm hibbāraʾăkā* recalls the reference to creation and naming of the primal human in Gen 5:2: *wayyiqrāʾ ʾet-šəmām ʾādām bəyôm hibbārʾām* 'and he called their name human *on the day they were created*'. We may also cite Gen 2:4a: *ʾēlleh tôlədôt haššāmayim wəhāʾāreṣ bəhibbārəʾām* 'These are the generations of the heavens and the earth, when they were created'. These linguistic similarities suggest the continuity of this passage with the tradition reflected in the opening chapters of Genesis, preserved by the priestly tradent.[293]

[293]Note that Gen 2:4a and 5:2 are generally assigned to P, with which Ezekiel would more than likely have been familiar.

The location of the primal human

In both oracles in Ezekiel 28, the primal human is situated within the realm of the divine. In 28:2 the primal human says, 'I dwell in the dwelling of the gods in the heart of the seas', and in 28:13 we read concerning him, 'You were in Eden, the garden of God'. The location in the divine realm was no doubt a feature of the mythic complex of ideas to which the audience was accustomed. The imagery appears in the Genesis account in the form of the garden of Eden, although there the significance of divine location may have been eclipsed by other concerns.

The imagery of the divine habitation is also present in the references to the *kərûb* in 28:14,16. It is relatively clear that the cherubim represent members of the divine court, but the precise relationship of the first human to these beings is not explicitly defined. Nevertheless, that a relationship exists is clear; it is expressed both through the location of the man in the garden/mountain and in the fact that he 'walks among stones of fire'. As we have seen, the ambiguity of the relationship of the primal human to the cherub is sufficient to cause considerable confusion over whether the primal human figure was *with* the cherub or *was* the cherub.

The degree to which the primal human "belongs" in this context of divine habitation is not explicit, but it is part of the power of this oracle that a clearly human protagonist is placed in the divine settings of garden and (holy) mountain.

The wisdom of the primal human

Mantic wisdom, or procuring of divine secrets, is a central facet in the two oracles to Tyre in Ezekiel 28.[294] In each, mantic wisdom manifests itself in Tyre's uncanny ability in international commerce. It is expressed differently, however, in each case.

In the oracle to the king of Tyre in vv. 11–19, the presence of mantic wisdom is seen in the priestly language used to describe the primal human. In v. 13 he possesses what is construed as the breast-

[294]Cf. my discussion of mantic wisdom in chapter 4.

piece of judgment, worn by the Israelite high priest. The breastpiece contained the Urim and Thummim, the sacred lots used by the priest to obtain divine discernment. Like the primal human in the first oracle, his wisdom is used to produce massive wealth, a situation which results in his eventual demise (v. 17).

Conflict and the primal human

The theme of conflict between God and the primordial being is perhaps the most recognizable of the elements in the text. In this oracle, the conflict stems from the figure's location and relation to divinity, particularly with respect to wisdom. The conflict is explicitly called 'sin' and is represented in the abuse of wisdom.

The misdeed leads to the primal human's expulsion from the divine abode, and death. The imagery is much like that found in Genesis 3. He is expelled ('profaned') from the holy mountain of God — a place also referred to as Eden, the garden of God — and driven out (literally 'made to perish from') the divine beings. As was stated at the beginning of this chapter, Ezek 28:11–19 reveals a structure that features a contrast of states or conditions and it is this contrast, reiterated several times over, that captures the essence of the conflict.

4. Is Job the Primal Human?

(Job 15:7–16)

JOB 15:7–8 PRESERVES ANOTHER EXAMPLE of the analogical use of the primal human concept. Once again, as in Ezek 28:11–19, the tradent's intent in conjuring up the image is not to present a historically-oriented narrative with aetiological overtones. Rather, it is to seize upon the essence of the primal human concept in a more timeless sense — its identity again coalesces with that of another protagonist, in this case Job.

How does the content or essence of the myth alluded to by Eliphaz relate to what we have observed in the opening chapters of Genesis and in Ezekiel 28? Close scrutiny of these texts reveals that far from being distinct and separate concepts, the protagonists are for all practical purposes one and the same, the archetypal ancestor, the mediating, interstitial being who links humanity with divinity.[295]

The primal human appears in the book of Job most clearly in 15:7–8 within the context of an argument that extends to v. 16. In this passage (vv. 7–16) the speaker Eliphaz mentions characteristic elements of the first man and discloses his understanding of the figure through the analogy he draws. Thus, again the information we receive about the first man is indirect, but Eliphaz presents his argument in a manner that reveals the close acquaintance of his audience with the myth.

The Text: Job 15:7–16

7 [a]הֲרִאישׁוֹן אָדָם[a] תִּוָּלֵד וְלִפְנֵי גְבָעוֹת[b] חוֹלָלְתָּ[c]׃
8 הַבְסוֹד[a] אֱלוֹהַ תִּשְׁמָע [b]וְתִגְרַע אֵלֶיךָ[b] חָכְמָה׃

---

[295]See the views of W. Bousset, *Die Religion des Judentums im späthellenistischen Zeitalter* (HNT 21; Tübingen: J. C. B. Mohr, 1926), and C. H. Kraeling, *Anthropos and Son of Man: A Study in the Religious Syncretism of the Hellenistic Orient* (New York: Columbia University Press, 1927), 151–54.

9 מַה־יָּדַעְתָּ וְלֹא נֵדָע תָּבִין וְלֹא־עִמָּנוּ הוּא׃
10 גַּם־שָׂב גַּם־יָשִׁישׁ בָּנוּ כַּבִּיר מֵאָבִיךָ יָמִים[a]׃
11 הַמְעַט[a] מִמְּךָ [b]תַּנְחֻמוֹת אֵל וְדָבָר לָאַט עִמָּךְ[c]׃
12 מַה־יִּקָּחֲךָ[a] לִבֶּךָ[b] וּמַה־[c]יִּרְזְמוּן עֵינֶיךָ[c]׃
13 [a]כִּי־תָשִׁיב[a] אֶל־אֵל רוּחֶךָ[b] וְהֹצֵאתָ מִפִּיךָ מִלִּין׃
14 מָה־אֱנוֹשׁ[a] כִּי־יִזְכֶּה וְכִי־יִצְדַּק יְלוּד אִשָּׁה׃
15 הֵן בִּקְדֹשָׁו[a] לֹא יַאֲמִין[b] וְשָׁמַיִם[c] לֹא־זַכּוּ בְעֵינָיו׃
16 אַף כִּי־נִתְעָב וְנֶאֱלָח אִישׁ־שֹׁתֶה כַמַּיִם עַוְלָה׃

7 Were you born the first human?
Were you brought forth before the hills?
8 Do you listen in the council of Eloah?
Do you seize wisdom for yourself?
9 What do you know that we do not know?
What do you understand that escapes us?
10 Both the gray and the aged are with us
greater than your father in years.
11 Are the consolations of El too small for you,
that he must speak a secret with you?
12 What has your heart taught you?
How your eyes flash!
13 For you turn your spirit against El,
and you send forth words from your mouth!
14 What is a man that he can be pure?
Or that he can be righteous — he who is born of a woman?
15 He trusts not even his holy ones,
And the heavens are not pure in his sight.
16 How much more the abominable and corrupt,
The man who drinks iniquity like water!

Notes:

7 [a–a] Syr *qdm ʾnš*, Tg הבקדמאה אדם: [b]Tg נולמתא, Syr *rmtʾ*. [c]Tg איתבריתא, Syr *ʾtbṭnt*, LXX επαγης.

8 [a] Tg *rz*, Syr *rʾz*, LXX συνταγμα. [b] LXX var. η συμβουλω σοι εχρησατο ο θεος. [c-c] Tg ואספי לוותך. Syr *wʾtglyt lk*. Vulg *et inferior te eius*.

11 LXX ολιγα ων ημαρτηκας μεμαστιγωσαι, μεγαλος υπερβαλλοντως λελαληκας. Syr *ʾzgr mnk lwḥmwhy dʾlhʾ* *wmll* *bnyḥ lwt npšk*. [a] Vulg *numquid grande*. [b] *lwḥmwhy*. [c] Tg חזי למיהוי גבר, Vulg *sed verba tua prava hoc prohibent*.

12 [a] LXX *ετολμησεν, Vulg *te elevat*, Syr *ʾtrym*. Tg ילפינך. [b] Tg רעיונך. [c–c] Vulg *magna cogitans adtonitos habes oculos*.

13 [a–a] Vulg *quid tumet* = Syr. [b] LXX Θυμον.

14 [a] Tg בר נש.

15 [a] LXX αγιων (pl.) = Vulg, Tg, Syr; Tg adds עילא. [b] Vulg *inmutabilis.* [c] Tg מרומא ואנגלי.

16 Syr *ʾpn mstlʾ wmtṭrp gbrʾnšʾ ʿwlʾ ʾyk myʾ.*

Verse 7. The plene spelling ראישון is found normally in the Samaritan Pentateuch and compares to variant spellings in Job 8:8 and Josh 21:10. Note the Targum's use of הבקדמאה (cf. Syr). The formation of the fetus gave rise to the LXX reading ἐπαγης (Dhorme, 210). Duhm argued reading גבהים for גבעות as a reference to the angels.

Verse 8. The Targum's translation of רז (and Syr rʾz) for סוד reflects the interpretation 'secret' (cf. Aq, Sym, Th; see Dhorme). LXX συνταγμα and Vulg *consilium* both reflect 'council'. The word סוד is used to express both, so it is original and could well be a double entendre.

Verse 10. This verse is not original to the LXX text. It is marked by an asterisk in both Jerome and Syro-hex and entered the Septuagintal tradition through Theodotion (noted in many catena mss.; Syro-hex; ms. 248); hence, the reading reflects the early received Hebrew text. The verse is omitted in the Coptic Sahidic.

Verse 11. The Greek of v. 11 defies retroversion to a Hebrew *Vorlage* anything like that found in MT. The verse may have a corruption at its source. This seems to be indicated by the presence of ολιγα ('few') and λελαληκας (pf. λαλεω, 'you have spoken'). The former most frequently translates Hebrew מעט and the latter is overwhelmingly the most common equivalent for דבר. We can be reasonably certain that תנחמות is original in MT's *Vorlage,* given its infrequency and the fact that it does appear again in Job 21:2 in a clearer context.

Verse 12. Some critics have followed LXX επηνεγκαν, reading ירומון 'to grow haughty' for ירזמון (see Gordis, who compares Ps 6:7 [intending Prov 6:17?]; and [wrongly!] 30:13, p. 161) Pope compares Prov 6:17; 30:13. Another possible emendation, followed here, is to consider the root רזם a metathesis for the correct רמז, which is found in Tg, Syr and some medieval Hebrew mss. Following Driver's observation that Aramaic רמז corresponds to διανευω in Greek, Dhorme pointed out the phrase διανευων οπθαλμω in Sir 27:22, and on this basis has accepted the hapax as it stands.

The Context of Eliphaz's Comments on the Primal Human

The speech of Eliphaz in chapter 15 comprises three distinct but related units: (1) an introduction in vv. 2–6 dealing with the general demeanor of Job, in particular the inappropriateness of his language; (2) an analogy drawn in vv. 7–8 and elaborated in 9–16, which brings the focus to the essential impurity of humanity; and (3) a litany of characteristics of the wicked in vv. 17–35, the balance of the chapter. Our concern in this chapter will focus on vv. 7–16. The first and most important reason for delimiting the passage as such is logical (as opposed to formal). 7–16 is a section which begins and ends with comparisons between beings who are "privileged" with respect to knowledge, on the one hand, and humans, on the other: in vv. 7 and 8 the privileged being is the primal human; in v. 15 the privileged beings are the 'holy ones'. Further, 15:7–8 establishes a *specific* comparison between the primal human and Job; 15:15–16 establishes a *general* comparison between the holy ones and humanity. A reading of vv. 7–16 yields the following principal themes:

I. The Wisdom of the Primal Human (vv. 7–10)
II. Conflict with the Divine (First Elaboration of 7–10; vv. 11–13)
III. The Uncleanness of Man (Second Elaboration of 7–10; vv. 14–16)

The following analysis will discuss the principal themes and ideas as they appear.

The Wisdom of the Primal Human (7–10)

It is the wisdom of the primal human that first leads Eliphaz to invoke him in this argument. As we have seen in the two previous chapters, the wisdom of the primal human was of considerable significance.

The primal human as "first human" (7)

הֲרִאישׁ֣וֹן אָ֭דָם תִּוָּלֵ֑ד
וְלִפְנֵ֖י גְבָע֣וֹת חוֹלָֽלְתָּ

Were you born the first human?
Were you brought forth before the hills?

Eliphaz begins the second part of his discourse in this chapter with a rhetorical question: 'Were you born the first human?'. The syntax of Eliphaz's question is interesting.[296] It might be argued that

[296] S. R. Driver and G. B. Gray (*A Critical and Exegetical Commentary on the Book of Job* [ICC; Edinburgh: T. & T. Clark, 1921], 95) read, 'Wast thou the first one born a man?'. They explain it as "the first to be a man." R. Gordis (*The Book of Job: Commentary, New Translation and Special Studies* [New York: Jewish Theological Seminary of America, 1978], 160) reads, 'Were you born the first among men?'.

Eliphaz is asking Job whether he is an incarnation of the first human. In support one could cite Mal 4:5 as an example of the notion that the spirit or character of a long dead hero may become reembodied in another living person. Job 15:7a reads, literally, 'Were you born the first of humankind?'. The sense of this verse is made clear by two other occurrences of the verb *yld* in the Niphal. In Eccl 4:14 the verb appears with the participle *rāš* in the accusative, 'born a poor man'.[297] Likewise, and even more interesting is Job 11:12: 'A stupid man will get understanding when a wild ass's colt is born a human' (*wəᶜayir pereʾ ʾādām yiwwālēd*).[298] The sense is to be born "in the state of," or literally as, something else.[299] NRSV 'Are you the first man that was born?' does not do justice to the Hebrew, but ultimately suffices as an appropriate translational equivalent. We shall see shortly that this is in fact the intent behind the question. The question is rhetorical; Eliphaz expects a negative answer, as he does to the second query, 'Were you brought forth before the hills?'.

In these words of Eliphaz, we learn that the first human was thought to have been born before the hills. The verbal root here is *ḥwl* which means 'to dance or writhe'. It is used in connection with birth imagery, denoting writhing in travail; and hence can render the meaning 'to bear or bring forth'. The meaning of the verb is clear in the parallelism here with *yld*, seen also in Isaiah 51:2, *habbîṭû ʾel-ʾabrāhām ʾăbîkem wəʾel-śārâ təḥôlelkem* 'Look to Abraham your father and to Sarah who bore you' (cf. Ps. 51:7, in which *ḥwl* is parallel to *yḥm*, 'to conceive'; and Prov 8:24,25). The first human is described as having come into existence through natural means, that is, through birth.

---

[297]*rāš* here may be construed as an adjective. Note, however, the substantival use in 2 Sam 12:3; Psalm 82; Eccl 5:7; Prov 10:4. See GKC §118p: "Participles...either after the verb...or before it...are to be regarded as expressing a state and not as being in apposition, since in the latter case they would have to take the article."

[298]On the textual and grammatical problems in this verse see Gordis, 124.

[299]E. Dhorme (*A Commentary on the Book of Job* [trans. Harold Knight; Camden, N. J.: Thomas Nelson and Sons, 1967], 210) considered the Niphal of ילד as retaining "its proper literal meaning (parallelism with חולל), whereas, in 11:12b, it had the meaning of assume the nature of some one, or become."

The poet chooses to use two words connoting natural birth, an image totally unlike that found in the Genesis accounts in which humankind is 'made', 'created', and 'formed'. The choice of words, however, does not appear to spring from the poet's own creativity. Rather his words belong to a stock of common conceptions regarding the creation. The image of natural birth and that of the forming of clay combine to express the creation of humankind in Mesopotamia, particularly in the actions of the mother-goddess.[300] This concept is nicely echoed in Ps 139:13: 'You formed (*qānîtā*) my inward parts; you knitted me (*təsukkēnî*) in my mother's womb'. This imagery appears also in Job 10: 'Your hands fashioned and made me' (*ᶜiṣṣəbûnî way-yaᶜăśûnî*; v. 8); 'Remember that you made me (*ᶜăśîtānî*) out of clay' (v. 9); 'Why did you bring me forth from the womb?' (v. 18).[301]

It is significant that this birth is said to have taken place 'before the hills', because in the P account of creation in Genesis, humankind is created *after* the waters have receded from the earth, allowing hills to appear.[302] It seems best, however to take the two cola of Job 15:7 together as simply a general expression of the remote mythical past.

Mountains and hills elsewhere signify antiquity. In the archaic prayer of Habakkuk, the prophet writes (3:6):

עָמַד ׀ וַיְמֹדֶד אֶרֶץ
רָאָה וַיַּתֵּר גּוֹיִם
וַיִּתְפֹּצְצוּ הַרְרֵי־עַד
שַׁחוּ גִּבְעוֹת עוֹלָם
הֲלִיכוֹת עוֹלָם לוֹ׃

He stood and measured the earth;
he looked and shook the nations
then *the eternal mountains* were scattered,

---

[300]E.g., *Gilg.* I ii 34–35; OB *Atr.* I 225f. See Tigay, *Evolution of the Gilgamesh Epic*, 195.

[301]Cf. also Jer 1:5.

[302]Dhorme considered this "an undisguised allusion to the discourse of Wisdom in Pr 8," and suggested considering each colon in Job 15:7 a reference to a different time. According to this understanding, the first colon refers to the creation of humankind "when he was still very close to Wisdom," while the second colon refers to an even earlier time, "before the creation." E. Dhorme, *A Commentary on the Book of Job*, 210. See my discussion of Prov 8:22–31 in chapter 6 below.

the *everlasting hills* sank low.
His ways were as of old.

The allusion here is to the march of the divine warrior in his history-making deliverance of the people of Israel. His ability to affect the very symbols of the earth's strength (the foundations of humanity's world) demonstrates for the writer the extent of his influence over the cosmos itself. The extreme antiquity of the mountains and hills serves to demonstrate the divine warrior's power — that which time had proved unmovable had been moved.[303]

Another passage which alludes to remote antiquity by referring to the mountains and hills and which lies very close to Job 15:7 is Ps 90:2:

בְּטֶרֶם ׀ הָרִים יֻלָּדוּ
וַתְּחוֹלֵל אֶרֶץ וְתֵבֵל
וּמֵעוֹלָם עַד־עוֹלָם אַתָּה אֵל׃

Before the mountains were *born*
and you *brought forth* the earth and the world,
and from eternity to eternity, you are God.

Here, *yld* and *ḥwl* appear together, this time applied to the mountains, the earth, and the inhabited world. This passage is strikingly similar to the language of Job 15:7, and it is difficult to consider them unrelated. Viewed alongside Job 15:8, 38:4–7 (when the sons of God rejoiced at the creation), and Prov 8:22–31 (to be discussed in chapter 6), Ps 90:2 contributes to a rough picture of the pre- (or early) cosmic situation, featuring Yahweh, the divine council of the *bənê ʾĕlôhîm*, and the one called the first human.[304] These passages also recall the observation of F. M. Cross, who pointed out how 'heaven', 'earth' (*ʾereṣ*), and 'world' (*tēbēl*) are used in Ps. 89:12 evoking "names of the old

---

[303]For further related citations and discussion of this passage, see T. Hiebert, *God of My Victory: The Ancient Hymn of Habakkuk*, (Atlanta: Scholars Press, 1986), 94–97. J. J. M. Roberts points out how this section (3:3–15) no longer refers to Yahweh's march as a past event, but as a "visionary experience" (*Nahum, Habakkuk, and Zephaniah: A Commentary*, [Louisville: Westminster/John Knox, 1991], 149). This does not affect the sense in which mountains and hills are invoked.

[304]See H. G. May, "The King in the Garden of Eden," 170. Cf. G. Hölscher, *Das Buch Hiob* (HAT 17; Tübingen: J. C. B. Mohr, 1937), 37–38.

gods."[305] Perhaps a similar allusion occurs here. The language is certainly suggestive of theogony, and at the least, makes abundantly clear that the first human spoken of by Eliphaz was born in the remote mythical past.

The council of God and the acquisition of wisdom (8)

הַבְסוֹד[a] אֱלוֹהַ תִּשְׁמָע
[b]וְתִגְרַע אֵלֶיךָ[b] חָכְמָה׃

Do you listen in the council of Eloah?
Do you seize wisdom for yourself?

Eliphaz continues his line of questioning by asking Job, 'Do you listen in the council of God?'. This imagery is consonant with that normally applied to prophecy. Generally speaking, the word *sôd*, translated both 'council' and 'counsel', is used in the Hebrew Bible to refer to a group or to that which transpires within a given group. When used to signify a group, it is used with reference both to humankind (e.g., Ezek 13:9) and to the divine realm (e.g., Ps 89:8). When used to signify that which transpires within a group, it appears sometimes with reference to private or closed conversation, again, within human groups or divine. One means of expressing the disclosure of this private or closed conversation is the verb *glh*.

The image of the divine council is well attested in the ancient Near East in Mesopotamia, Ugarit, Phoenicia, and Israel. The general purpose of these assemblies was to decide the fates of the cosmos and its inhabitants. Mesopotamian literature features the designation *puḫur ilāni* 'assembly of the gods'. In the west, Ugaritic attests *pḫr ʾilm* and Phoenician, *mpḥrt ʾl gbl qdšm*, 'the assembly of the holy gods of Byblos'.[306] Hebrew uses a variety of terms for the council,

---

[305]F. M. Cross, *Canaanite Myth and Hebrew Epic*, 160; cf. *idem*, "'The Olden Gods' in Ancient Near Eastern Creation Myths," in ***Magnalia Dei, The Mighty Acts of God: Essays on the Bible and Archaeology in Memory of G. Ernest Wright*** (ed. F. M. Cross, W. E. Lemke, and P. D. Miller, Jr.; Garden City, N.Y.: Doubleday, 1976). Note Josh 5:13–15; 10:12b–13a; Judg 5:20; Ps 148:2,3 as examples of the heavenly bodies being depicted as members of the entourage of Yahweh. Note also invocations of heaven in earth in passages like Deut 32:1.

[306]E. T. Mullen, "Divine Assembly," *ABD* 2.214.

some of which appear in Ugaritic as well, including *ᶜēdâ* (Ps 82:1) and *dôr* (Amos 8:14).[307] Words used in Hebrew for the council which have not been attested in Ugaritic are *qāhāl* and *sôd*.[308] *sôd,* used here in 15:8, refers to the council of God — a group closed to humans simply by virtue of its designation; hence, a *secret* council.[309] The relevance of this secrecy for our passage is that there are nevertheless situations in which humans are related to the divine council (*sôd*). The clearest examples concern the prophets.[310] In Amos 3:7 we are told: 'Surely Yahweh God does nothing without revealing his *sôd* (*kî* *ᵓim-gālâ sôdô*) to his servants the prophets'.

The significance of the imagery of prophecy in Job 15 is even more clearly evident in the language that surrounds the word *sôd* in this passage. Although Gordis interprets the *bə-* of the phrase *šāmaᶜ bə-* to suggest 'listening in, eavesdropping', there are no other apparent examples of this nuance (i.e., unauthorized listening) of the phrase in the Hebrew Bible.[311] In fact, *šāmaᶜ bə-* appears with the simple meaning of 'to hear about' in Ps 92:12.[312] I should like to suggest that a more appropriate background of the imagery may be found in Jer 23:18:

---

[307]See also Pss 14:5; 49:20; 73:15 (Mullen, "Divine Assembly," lists additional examples). The use of *ᶜdt* in this sense in Ugaritic occurs in *CTU* 1.15.II.7,11. Among the numerous examples of *dôr* in this sense in Ugaritic are *CTU* 1.65.2; 1.40.7, 25, 33–34, 42; see Mullen, *The Divine Council in Canaanite and Early Hebrew Literature* (Harvard Semitic Monographs; Chico, Calif.: Scholars Press, 1980), 118, for additional examples.

[308]For *qāhāl*, Mullen cites Ps 89:6; for *sôd* he cites Ps 89:8; Jer 23:18 (cf. v. 22); and Job 15:8.

[309]On the use of *sôd*, see Fabry, *TWAT* 5.776.

[310]There are other passages in which *sôd* has traditionally been interpreted to refer to intimacy or friendship with God (see BDB, 691). This intimacy is extended to the righteous in Ps 25:14.

[311]Gordis offers no support for this interpretation. If anything, *šāmaᶜ bə-* suggests the notion of obedience, especially in conjunction with *qôl*, 'voice'. BDB classified *šmᶜ* with other verbs of "sensible perception" that are used with the preposition *bə-*, each of which tends to take the preposition to emphasize attentiveness (90). Thus, the basic distinction between *hear* and *listen* is underscored.

[312]BDB (1033) may be correct in suggesting that *šāmaᶜ bə-* in this verse has been influenced by *hibbîṭ bə-* of the preceding colon.

כִּי מִי עָמַד בְּסוֹד יְהוָה וְיֵרֶא וְיִשְׁמַע אֶת־דְּבָרוֹ
מִי־הִקְשִׁיב דְּבָרִי [דְּבָרוֹ] וַיִּשְׁמָע

For who has stood in the council of Yahweh and perceived *and heard* his word?
Who has been attentive to [his] word[313] *and has listened*?

Here, *sôd* is used to designate the divine council, and it is used with the verb *šmᶜ*. The imagery is that of listening *within*; one stands within the council of Yahweh and perceives and hears his *dābār*. The evidence suggests that Job 15:8 reflects the normal understanding of the divine council and the dispatching of the divine decree. The language itself does not convey the idea of stealing or deception, as scholars such as Gunkel have imagined (see below). Ancient Israel saw nothing unusual in the participation of certain human beings in the divine realm, but that participation was limited to certain offices, in the case of Jeremiah, that of prophecy.

The interpretation of Eliphaz's next question has been of central importance in establishing the nature and identity of the first human spoken of here. Gunkel's interpretation at the turn of the century typifies how many commentators have understood it:

> Daß auch 8 zu dieser anzunehmenden Anspielung gehört, zeigt besonders das Wort תִּגְרַע »du rissest an dich«, »unterschlugest«; Hiob wird mit einem Wesen verglichen, das wider Gottes Willen in seinen Rat eindringt und sich Weisheit raubt. Wie aber soll der Dichter auf diese Vorstellung gekommen sein, wenn man nicht eben dies vom Urmenschen erzählte? — Der vorausgesetzte Mythus berichtete also, daß der »erste Mensch«, der vor der Welt gekreißt worden war, im himmlischen Rat zugehört und sich so Weisheit gestohlen hat.[314]

Here 'to seize for oneself' and 'to withhold' become 'to steal'. Gunkel

---

[313]Reading Qere 'his word' (*dĕbārô*) for דברי. The *yod* of דברי may have arisen in a graphic confusion with an original *wāw*.

[314]"The word *tigraᶜ* 'you seized for yourself', 'you withheld', in particular shows that v. 8 belongs to this proposed allusion: Job is compared to a creature, which penetrated against God's will in his council and stole for himself wisdom. But how should the poet have come upon this idea when this was not even told about the primal man? The assumed myth therefore reports that the 'first man' who was brought forth before the world, listened in the heavenly council and thus stole wisdom for himself." H. Gunkel, *Genesis*, 34.

was influenced by the Greek myth of Prometheus, who stole fire from the gods. It is not impossible that Prometheus is somehow related to this figure; but it is difficult to get a meaning 'steal' from the verb *grᶜ*. It is more likely that something a little more subtle is intended. At its root, *grᶜ* seems to carry the meaning 'to take (from) something'. The verb is used in a variety of ways. It connotes 'to trim' a beard (Jer 48:37); and 'to take away' (of the commandments of Yahweh, Deut 4:2; 13:1; cf. Num 36:3 of tribal inheritance), which may also be construed as 'diminishing' (a collective). In Exod 21:10 it appears to bear the meaning 'diminish', although one might make a case for 'withhold' in this context. In Job 15:4, *grᶜ* appears with the direct object *sîḥā*, 'meditation' and is parallel to the verb *prr* 'to annul' which takes *yirʾâ* 'reverence, fear' as its direct object (*prr* is most often used to express breaking a covenant or rendering counsel ineffectual). Here in v. 8, the sense seems to be 'doing away with', or more literally, 'taking away' meditation. One might still argue for 'detracting from'. It is clear that *grᶜ* implies taking either *some* or *all* of a thing.[315] While the precise meaning of the verb is difficult to grasp, given what we know of *grᶜ*, I believe it is safest to translate 'to seize'. A more nuanced meaning is, in my judgment, unwarranted.

Regarding the use of the imperfect verbs *tišmaᶜ* and *tigraᶜ*, Driver and Gray observed that either it is a depiction of the past "alluding to the particular divine council [cf. Gen 1:26] in which the plan of creation was revealed" or it "indicates recurrency...'art thou wont to be a listener.'" They conclude that in either case *tigraᶜ* may not have been intended to be "coordinate" but "consecutive," that is, "did (or do) you listen, and so draw to yourself."[316] This interpretation does not, however, shed much light on the *intent* of the allusion here. Perhaps more noteworthy from a grammatical standpoint is that the first colon of both vv. 7 and 8 reflects the construction: question marker + verbal object + imperfect verb and subject, which suggests

---

[315]Dhorme asserted that *grᶜ* 'to diminish, do away with' also means 'to claim for oneself' or 'attract to one's self', and with *ʾēleykā* he arrives at the translation 'have a monopoly of'. *A Commentary on the Book of Job*, 211. See also BDB, 175.

[316]Driver and Gray, *A Critical and Exegetical Commentary on the Book of Job*, 96.

that the two verses were in fact intended to be taken together. That is, the *sôd* is presented here with reference to the primal human.

The meaning of "wisdom" in 15:8

It is necessary to define the "wisdom" of which Eliphaz speaks in v. 8, primarily because it constitutes *the* defining characteristic of the first human to whom he refers. The easy solution would be to consider it a general, all-encompassing reference to wisdom. Is a more specific application intended here?

It must be stated at the outset that the term *ḥokmâ* is not easily apprehended. The root *ḥkm* is shrouded in ambiguity as it seems to exhibit a broad semantic range from a specific technical knowledge to general intelligence. Aside from the definitions of wisdom as a corpus of literature, a world-view, or a tradition, the center of wisdom seems to have a very practical orientation.[317]

Of those who have considered wisdom to represent a particular view of reality, Crenshaw understands it to be, among other things, a fundamental point of orientation which is concerned with "what is good" for humanity; the answers to this quest are to be found in experience. He continues:

> That world view assumes a universe in the deepest and richest sense of the word. The one God embedded truth within all of reality. The human responsibility is to search for that insight and thus to learn to live in harmony with the cosmos. In a sense the sage knows the right time for a specific word or deed. It follows that optimism lies at the center of wisdom. This confidence in the world and human potential gives rise to profound skepticism, of course, but such a heart-rending cry bears eloquent testimony to a grand vision of what ought to be, a vision that persists even though despair has overwhelmed the sage.

In her study of Mesopotamian wisdom, S. Denning-Bolle notes:

> We intuitively make distinctions, assuming that these distinctions are essential. Nevertheless, our mental habits are not always illuminating.... For example, it seems natural for us to differentiate two types of wis-

---

[317]For discussions of the definition of wisdom, see Crenshaw, *Old Testament Wisdom* (Atlanta: John Knox, 1981), 11–25; "Wisdom in the OT" *IDBSup* 952–56; R. Murphy, "Wisdom in the OT" *ABD* 6.920.

> dom: first, a practical down-to-earth wisdom where advice is offered on how to conduct one's life; and second, a more abstract type of wisdom where one finds an enquiring into the enigmas of life and an attempt to answer some of these enigmas. This differentiation was not so apparent to the ancient Mesopotamian.... Our texts demonstrate a great deal of overlapping with respect to the two categories we have constructed.[318]

A study of the Hebrew term *ḥokmâ* reveals that the term possesses no intrinsic moral quality. It is not connected exclusively with any type of moral or ethical action, and hence, cannot uncritically be equated with the notion "good."[319] At its root, before anything else, lies the concept of knowledge, or better, *effective knowledge*. It is effective knowledge because it is the knowledge of the laws and constituents of the cosmos. The important part here is the laws. Discovery and knowledge of natural laws (order and causation) leads to the practical articulation or application in technology. The ancient mind had no reason to separate natural and spiritual causes; all phenomena operated in accordance with the cosmos — the order of which was not always possible to trace. This order may be described as a state, but is often portrayed as a possession; the phrase 'full of wisdom' (1 Kgs 7:14; Ezek 28:12) hints that one possesses it in degrees. Therefore, the practical outcome of wisdom is potentiality. When one has wisdom, one is free to apply it for "good" or for "evil" (cf. Jonadab in 2 Sam 13:3–5). At its root, all phenomena come under its domain. Individuals can possess it, however, with regard to specific phenomena. In this regard, it is not surprising that the theme of creation is inextricably woven with the concept of wisdom.[320]

ḥokmâ is both cosmic knowledge and its logical correlate, power. To have *ḥokmâ* is to have power over something. In Jer 9:16,17, we read of women who have the power to evoke grief: 'Call for the mourning women to come; send for the skillful women (*ḥăkāmôt*) to

[318]S. Denning-Bolle, *Wisdom in Akkadian Literature: Expression, Instruction, Dialogue* (Leiden: Ex Oriente Lux, 1992), 31–32.

[319]It is true that one must also consider the historical movement of the term in time; different epochs may reveal different understandings of wisdom. Thus, a synchronic study has its limitations as well.

[320]According to Murphy ("Wisdom in the OT," 924), "wisdom and creation are mirror images of each other."

come; let them make haste and raise a wailing over us that our eyes may run down with tears, and our eyelids gush with water'.[321]

Authority, particularly with respect to kingship was to a large degree based on the possession of *ḥokmâ*. Possessing *ḥokmâ* was not simply considered a good thing, or requisite; it was what enabled one's reign to continue and thrive.[322] The encyclopedic knowledge associated with *ḥokmâ* is not simple book knowledge, of value solely because it is interesting; rather it is a further expression of one's intimacy with the cosmos. The more one possesses *ḥokmâ*, the more its manifold fruits are evident. The author of the pseudepigraphic Wisdom of Solomon expressed this potentiality centuries later in 7:15–22:

> May God grant me to speak with judgment,
> and to have thoughts worthy of what I have received;
> for he is the guide even of wisdom and the corrector of the wise.
> For both we and our words are in his hand,
> as are all understanding and skill in crafts.
> For it is he who gave me unerring knowledge of what exists,
> to know the structure of the world and the activity of the elements;
> the beginning and end and middle of times,
> the alternations of the solstices and the changes of the seasons,
> the cycles of the year and the constellations of the stars,
> the natures of animals and the tempers of wild animals,
> the powers of spirits and the thoughts of human beings,
> the varieties of plants and the virtues of roots;
> I learned both what is secret and what is manifest,
> for wisdom, the fashioner of all things, taught me.[323]

This knowledge is power and it can translate to one's good or another's evil. Wisdom enables one to have power over a thing or

---

[321]H.-P. Müller (*TDOT* 4.378) classifies this passage as an example of "magical wisdom." The NRSV 'skillful' is not a bad translation, so long as it is taken without dividing secular and sacred. Another similar example is Ezek 27:8,9.

[322]Solomon's successes are tied to the degree to which he possessed wisdom *ḥŏkmat ʾĕlôhîm*. He possessed it more than any other; therefore he was opposed by none, and the land was at peace.

[323]The verses that follow show wisdom to be intrinsically good, but the passage is late and comes after the establishment of the identification of wisdom with Torah, and wisdom with the 'fear of Yahweh'; The identification of wisdom and Torah could yield no other conclusion but that wisdom is an inherently good thing.

another person. Jonadab is called *ʾîš ḥokmâ,* which is lost in NRSV 'a very crafty man' (2 Sam 13:3). The writer simply wished to point out why Jonadab was the one to offer advice. The description lacks a value judgment. The wise person offers counsel because he or she understands the workings of the cosmos and can get a person what he or she wants, so long as it within the realm of cosmic principles.

The examples cited above point to a very practical or technical orientation of wisdom. One finds further corroboration in the words used to express wisdom in the cultures of the ancient Near East, often having to do with skill (Akkadian *nēmequ*, etc.). This practical orientation would not necessarily lie in the conscious thoughts of those using the term *ḥokmâ,* and although it is quite difficult to sort out, one may suggest certain types or uses of the term. H.-P. Müller perceives four types of wisdom in the Hebrew Bible. In addition to a "non specific usage," Müller identifies court wisdom, mantic wisdom, magical wisdom, and wisdom that pertains to artisans.[324] Very close to the practical orientation of wisdom discussed above is what has been termed mantic wisdom.[325] Of the possible types of wisdom which may be perceived in the Hebrew Bible, it is this so-called mantic wisdom which lies closest to the ideas expressed in vv. 8 through 10.

Balaam's oracle in Numbers 24:15–19 provides an example of manticism and relates images not unlike those found in Job 15:7–9. In v. 16 we read:

נְאֻם שֹׁמֵעַ אִמְרֵי־אֵל וְיֹדֵעַ דַּעַת עֶלְיוֹן מַחֲזֵה שַׁדַּי יֶחֱזֶה נֹפֵל וּגְלוּי עֵינָיִם׃

> The oracle of him who *hears the words of God* and *knows the knowledge of the Most High,* who sees the vision of the Almighty, falling down, but having his eyes uncovered.

Here, hearing is used in an explicit, technical sense relative to oracle-giving and prophecy. A claim to know the knowledge of the Most

[324]Müller, *TDOT* 4.373–78. Note that Müller does not draw a hard distinction between mantic and magical wisdom: "In practice, magic and manticism are closely related" (378).

[325]I define mantic wisdom as the procuring of divine knowledge. It is exemplified in prophecy (cf. Greek *mantikos* 'prophet') and in the practice of divination.

High is not a casual statement. Rather, it is a lofty claim which highlighted Balaam's distinctiveness and proclaimed the integrity of the oracle.[326] Against this background, the significance to Eliphaz of the first human who listened in the divine council and his question, 'What do you know that we do not know?', become clearer.

The emphasis on mantic wisdom is evident in Job, specifically in the explicit references to dreams or visions.[327] In 4:12–19 Eliphaz relates a revelatory experience, in which the truth he again proclaimed to Job in chapter 15 was announced to him (4:17–19). Similarly in 33:15,16, Elihu appeals to dream revelation as legitimation of his right to speak and of his claim to wisdom.

Eliphaz says of Job in 15:5 that he 'chooses the tongue of the crafty'. This is the speaker's second time posing the example of the crafty. In 5:12 he says concerning Yahweh:

מֵפֵר מַחְשְׁבוֹת עֲרוּמִים
וְלֹא־תַעֲשֶׂינָה יְדֵיהֶם תּוּשִׁיָּה׃
לֹכֵד חֲכָמִים בְּעָרְמָם
וַעֲצַת נִפְתָּלִים נִמְהָרָה׃

He frustrates the devices of the crafty,
so that their hands achieve no success.
He takes the *wise in their own craftiness*,
and the schemes of the wily are brought to a quick end.

It becomes clear quickly in reading the book of Job that, as elsewhere in the Hebrew Bible, wisdom may be construed as positive or negative. It is a term laden with ambiguity and potentiality. Still, we may observe concerning its use with reference to the primal human that its connection to the divine realm is emphasized; wisdom belongs to the *sôd ᵓĕlôaʰ*. The fact that the primal human possesses it is unusual; moreover, it is part of what defines him as special and appropriate for the analogy. The fact that Eliphaz relates that the first human 'seized' wisdom recalls the Genesis 2–3 account. There, we are told regarding the tree of knowing good and evil — a tree that was desired to make

[326] Num 25:15. Müller cites this passage as an example of the use of the root *ydᶜ* as mantic knowledge (*TDOT* 4.376–78).

[327] Note how dream and vision are in close connection in 20:8.

one wise (*ləhaśkîl*) — that 'the woman *took* (*tiqqaḥ*) of its fruit and ate'.

Wisdom, knowledge, and primordiality (9–10)

9 מַה־יָּדַעְתָּ וְלֹא נֵדָע
תָּבִין וְֽלֹא־עִמָּנוּ הֽוּא
10 גַּם־שָׂב גַּם־יָשִׁישׁ בָּנוּ
כַּבִּיר מֵאָבִיךָ יָמִים

9 What do you know that we do not know?
What do you understand that escapes us?
10 Both the gray and the aged are with us,
greater than your father in years.

Job's knowledge versus the comforters' (9)

Wisdom in the second colon of v. 8 provides an appropriate segue into vv. 9 and 10 where the intent of Eliphaz's allusion begins to emerge. Eliphaz's question in v. 9, 'What do you know that we do not know? What do you understand that escapes us?', pits Job against the comforters in terms of wisdom. The verbs *ydᶜ* and *byn* are commonly paired elements, found together often in the Hebrew Bible (e.g., Isa 6:9; 44:18; Prov 24:12; Job 14:21).[328] A survey of these occurrences suggests a reference to the concepts of knowing and understanding in their most general sense. Thus, there is little in the language itself here that recommends positing a deeper meaning. Still, the immediate context of this verse suggests more than an incidental use of *ydᶜ* and *byn*. Its significance lies within the structure of the argument. In its context within vv. 7–10, v. 9 casts a great deal of light on these verses as the nexus of the ideas expressed in the preceding two verses (7 and 8) and following verse (10). In this section, vv. 7 and 8 express Job's perceived claim to higher knowledge; v. 9 questions or refutes it; v. 10 expresses the comforters' claim to higher knowledge. In vv. 7–8 and in v. 10, antiquity is invoked as a (or *the*) criterion of authority. In the first

[328]H. Ringgren notes the fact that the pairing of *ydᶜ* and *byn* occurs "in very different contexts" (*TDOT* 2.101).

instance, it is accomplished by means of explicit reference to the first human; in the second it is by reference to the ancients. It is clear throughout the speeches that the three friends attempt to comfort Job not through empathy, but through enlightenment and that toward this end, the claiming of wisdom takes on an importance of the first order. The disjuncture between the conversants and their ancient foils requires that these ideas be considered more closely.

Elders and ancestors (10)

In v. 10, Eliphaz's claim to authoritative knowledge is based on his association with those whom he refers to as the *śāb* and the *yāšîš*, whom he juxtaposes with Job's father. What do these terms signify? *śāb* comes from the root *śyb* which essentially means 'to be gray'; it is a relatively small logical step to the concept 'aged' (cf. Prov 20:29). It is related to Ugaritic *šyb* which expresses both grayness and old age, and Aramaic *syb* which means 'to be old'.

Lev 19:32 states the dictum, 'You shall rise before the aged (*śēbâ*, hoary head) and defer to the old man (*zāqēn*) and you shall fear your God...'.Verse 32 could have been intended simply to stand alone, an exhortation to give respect to old people; and because chapter 19 comprises so many loosely connected precepts, it would be somewhat tenuous to consider the immediate context of the verse.[329] It might have been understood as belonging to the prohibitions against exploitation. However, one might conjecture that it is best to understand it together with the preceding verse, 'Do not turn to mediums or wizards; do not seek them out, to be defiled by them...'. In this sense, it is the positive commandment, alternative to the previously stated prohibition.[330]

yāšîš, 'aged' is conjectured to come from the root *yšš*, a root which in Arabic denotes weakness or impotence. It is unique to the book of Job, where it appears in four passages as an adjective or substantive. In all but one instance in Job, it appears within the context of

[329]See, for example, Noth's comments on vv. 26-31 (*Leviticus: A Commentary* [Philadelphia: Westminster, 1977], 143).

[330]It is perhaps not insignificant that the verse closes with 'and you shall fear your God', cf. 'the beginning of wisdom'.

a discussion regarding wisdom. We will briefly survey these occurrences to see what insights they might yield with regard to chapter 15.

The first appearance of the term *yāšîš* in Job is in 12:12. Here, wisdom is said to be their possession:

בִּישִׁישִׁים חָכְמָה
וְאֹרֶךְ יָמִים תְּבוּנָה

Wisdom is with the aged,
and understanding with length of days.

Four elements of this maxim appear in 15:8–10: wisdom (*ḥokmâ*, v. 8), understanding (*təbûnâ*, cf. v. 9), the term *yāšîš* (v. 10) and the reference to length of days (*ʾōrek yāmîm*, cf. v. 10 *kabbîr...yāmîm*). Although the intent of the maxim is clear, however, this is followed by the somewhat enigmatic statement in 12:13, 'with him are wisdom and might (*gəbûrâ*); he has counsel (*ʿēṣâ*) and understanding'. The antecedent subject ostensibly is Yahweh, although a few medieval manuscripts read *ʾĕlōᵃh*, God. What is important here, it seems, is the addition of *gəbûrâ* and *ʿēṣâ*. One way to interpret v. 13 is as a complete negation of what may be reckoned as a popular saying in v. 12. In other words, "*you have heard it said that...but* I *tell you*...." That this saying was commonly known and open to question is clear from 32:6–9. JPS offers a similar understanding in rendering 12:12 an unmarked question, which is in turn answered by v. 13. Another way to make sense of these two verses, and one which I am inclined to accept, is to see v. 13 as simply adding to v. 12. In other words, the 'aged' and the 'long of days' have wisdom and understanding, but Yahweh or God has *gəbûrâ* and *ʿēṣâ* beyond this. The lack of any literary cue detracts from the force of arguing a complete negation; however, the verses are intentionally placed side by side for consideration by the audience. Yahweh (or God) is placed alongside the aged — the human channels of wisdom — and is thereby compared and contrasted.

The next occurrence of *yāšîš* is 15:10. It is difficult to ascertain the meaning of the second colon of v. 10 in its place following the first. The ambiguity arises in part out of the choice of *bānû* to express the relationship between the comforters on the one hand and the *śāb* and *yāšîš* on the other. The preposition *bə-* reveals little in this respect. Of the many ways in which the preposition is used, the most reasonable interpretation in this case is 'among' or 'with'. Often in the Hebrew Bible, the preposition *bə-* denotes presence of a given subject among additional members of its class, or one among more of the same. To relate a subject to a group by means of the preposition *bə-* can be to identify that subject as a member of the group. Thus, in 2 Sam 15:31, Ahithophel is *baqqōšərîm*, 'among' the conspirators, or 'one of the conspirators'. Another well known example is Amos 1:1 which describes Amos as being *bannōqədîm mittəqôaᶜ*, 'among' the sheepherders from Tekoa. Of course, context is what determines meaning, and quite frequently *bə-* denotes the sense of simple accompaniment, without equating the two. There are many examples of this use in verbal clauses (e.g., 1 Kgs 10:2; Isa 8:16). More instructive for the purposes of Job 15:10 would be examples of this use of *bə-* in a *verbless* clause where the potential ambiguity (of the association) is greater.[331] The only clear example appears to be 1 Kgs 19:19, which says of Elisha: *wəhûᵓ bišnêm heᶜāśār* 'he was with the twelfth (yoke of oxen)'.[332] There are apparently no examples in the Hebrew Bible of the preposition *bə-* denoting 'with' (association) in a verbless clause where both subjects are human. The standard way of expressing 'with' is by means of the preposition *ᶜim* or *bətôk*. Thus, if the language itself indicates anything in Job 15:10, it suggests the interpretation 'among' in the sense of identification (or identification through association).

There are other instances of *yāšîš* in Job which further aid in clarifying this use by Eliphaz. Job 29:8, 'The young men (*nəᶜārîm)* saw me and withdrew, and the aged (*yəšîšîm*) rose and stood', places the *yəšîšîm* in opposition to the *nəᶜārîm*, a generic term for youth. This juxtaposing speaks against understanding *yāšîš* as some sort of technical term. One might see in this an expression of range in age, status,

[331]Grammarians have traditionally understood that the noun clause in which the predicate is a substantive (as opposed to an adjective) frequently stresses the identity of subject and predicate (GKC §141c). GKC classifies constructions with prepositions as adverbial and, thus, understands the prepositional phrase to be "the equivalent of a noun-idea" (§141b).

[332]A related example of this use is the phrase *ᶜēṣ bəlaḥmô* 'tree with its fruit' (Jer 11:19).

and authority; that is, from the least empowered male members of society (boys) to the most powerful. It does appear to be a reiteration of the idea that Job was the greatest of the sons of the east; for even those to whom society ascribed wisdom (the elders) stood (presumably in honor) in his presence (perhaps best understood alongside Lev 19:32, mentioned above: 'You shall rise before the aged and defer to the old man, and you shall fear your God...').

Vocabulary and phrasings similar to 15:10 also occur in the introduction to the Elihu speech in Job 32:6–9:

> [6]וַיַּעַן ׀ אֱלִיהוּא בֶן־בַּרַכְאֵל הַבּוּזִי וַיֹּאמַר צָעִיר אֲנִי לְיָמִים וְאַתֶּם
> יְשִׁישִׁים עַל־כֵּן זָחַלְתִּי וָאִירָא ׀ מֵחַוֺּת דֵּעִי אֶתְכֶם׃ [7]אָמַרְתִּי יָמִים
> יְדַבֵּרוּ וְרֹב שָׁנִים יֹדִיעוּ חָכְמָה׃ [8]אָכֵן רוּחַ־הִיא באֱנוֹשׁ וְנִשְׁמַת שַׁדַּי
> תְּבִינֵם׃ [9]לֹא־רַבִּים יֶחְכָּמוּ וּזְקֵנִים יָבִינוּ מִשְׁפָּט׃

> [6]And Elihu the son of Bar'achel the Buzite answered: "I am young in years, and you are aged; therefore I was timid and afraid to declare my opinion to you.[7] I said, 'Let years speak, and many years teach wisdom.' [8]But it is the spirit in a man, the breath of the Almighty, that makes him understand. [9]It is not the great (in years) who are wise, nor the aged who understand what is right."

The phrase ***kabbîr mēʾābîkā yāmîm*** in 15:10, 'greater than your father in years', which is reflected in ***ʾōrek yāmîm*** 'length of years' in 12:12, is echoed here in ***yāmîm*** 'days' and ***rōb šānîm*** 'many years' of 32:7.[333] Through the vehicle of Elihu, the author identifies each of the three comforters (and perhaps Job) as ***yāšîš***. His intent is to turn the maxim equating old age with wisdom on its head. Between 32:6 and 32:9 it is clear that the author uses the term ***yāšîš*** synonymously with ***rab*** and ***zāqēn***.

On a related point, just as Lev 19:32 made ***śēbâ*** parallel to ***zāqēn***, 2 Chr 36:17 pairs ***yāšēš*** (cf. ***yāšîš***) with ***zāqēn***. In this verse, the two are opposed to ***bāḥûr*** 'choice young man' and ***bətûlâ*** 'virgin, maiden'. This survey of terms lends support to the conclusion that the comforters are identified (by association) with the ***śāb*** and ***yāšîš*** in Job 15:10.

The text most instructive for understanding the reference to antiquity in Job 15 is found in the speeches of the divine theophany in

---

[333]Cf. also the opposing phrase ***ṣāʿîr... lĕyāmîm*** 'young in years'.

38–41. This section exemplifies the type of encyclopedic knowledge that the sage possessed and that was both a sign and an expression of his or her wisdom; that is, not only does it mark a person as wise, such statements about the cosmos were a revelation of things not known. In 38:21 Yahweh closes a series of questions with the statement:

יָדַעְתָּ כִּי־אָז תִּוָּלֵד
וּמִסְפַּר יָמֶיךָ רַבִּים

You know, for you were born then,
and the number of your days is great.

Three things stand out in this apparent expression of sarcasm. First, the divine questioner implies that Job would in fact know the answers to his queries, had he been alive when the cosmos was established. Second, Yahweh's statement is replete with the language and imagery found in the speech of Eliphaz in chapter 15. The clause 'You know' (*yādaᶜtā*) is consonant with the question of Eliphaz, 'What do you know?' (*ma(h)-yādaᶜtā*); both are linked to a sarcastic reference to Job's birth (*hărîʾšôn ʾādām tiwwālēd*, 15:7; *ʾāz tiwwālēd*, 38:21). The clause *mispar yāmeykā rabbîm* recalls *kabbîr mēʾābîkā yāmîm* of 15:10. Third, the language expresses with unquestioned clarity the concept of *old with respect to the past*. To ascribe 'many days' to someone is not to call him *old man*, but *man of old*. Still, Yahweh's statement is facetious, for we are not led in any way by the author to identify Job with the first human here. It is clear, however, that Yahweh's statement implies that the answers are somehow bound up with creation. The wisdom that Job would possess would be of use in its mantic aspect; that is, the specific answers to Job's particular situation lie somewhere in the patterns and decisions of creation. If Job had been present at the creation, he would possess the wisdom that would explain his sufferings. The author once again tells Job, but this time from the mouth of Yahweh himself, that he is not the first human. Job was not born in the primordial age and, hence, did not know the reasons for his sufferings—this, despite the fact that he believed the problem to lie with God and not with himself (cf. 11:4, 16:17; 33:9). In the mind of the poet, as expressed in chapter 15, the privileged knowledge that yields the answers to the intractable puzzles of life comes from the primordium. Job indeed did not know, *but one born then would;* that one is the 'first human'.[334]

That the locus of wisdom is the primordium is expressed in the climax of the wisdom poem of chapter 28:

23 אֱלֹהִים הֵבִין דַּרְכָּהּ
וְהוּא יָדַע אֶת־מְקוֹמָהּ
24 כִּי־הוּא לִקְצוֹת־הָאָרֶץ יַבִּיט
תַּחַת כָּל־הַשָּׁמַיִם יִרְאֶה
25 לַעֲשׂוֹת לָרוּחַ מִשְׁקָל
וּמַיִם תִּכֵּן בְּמִדָּה
26 בַּעֲשֹׂתוֹ לַמָּטָר חֹק
וְדֶרֶךְ לַחֲזִיז קֹלוֹת
27 אָז רָאָהּ וַיְסַפְּרָהּ
הֱכִינָהּ וְגַם־חֲקָרָהּ

23 God understands the way to it,
and he knows its place.
24 For he looks to the ends of the earth
and sees everything under the heavens.
25 When he gave to the wind its weight,
and meted out the waters by measure;
26 when he made a decree for the rain,
and a way for the lightning of the thunder
27 then he saw it and declared it ;
he established it, and searched it out....[335]

Wisdom is a divine characteristic mediated to humankind through the ancients. The significance of the primal human and the issue of antiquity in this passage may be made clear by the Mesopotamian interest in antediluvian knowledge. Several texts reveal the high value placed on this unique "higher" form of knowledge. Most significantly for our purposes, Adapa was noted to have recorded esoteric knowl-

---

[334]Note that May suggested the relevance of Job 38:4–7 in "The King in the Garden of Eden" p. 170. See also R. N. Whybray, ***The Heavenly Counsellor in Isaiah xl 13–14: A Study of the Sources of the Theology of Deutero-Isaiah*** (Society for Old Testament Study Monograph Series; Cambridge: Cambridge University Press, 1971), 55.

[335]For a different translation and interpretation, see Michael D. Coogan, "The Goddess Wisdom—'Where Can She Be Found?'" ***Ki Baruch Hu: Ancient Near Eastern, Biblical, and Judaic Studies in Honor of Baruch A. Levine*** (ed. R. Chazan, W. Hallo, and L. Schiffman; Winona Lake, Ind.: Eisenbrauns, 1999), 203–9.

edge 'from before the flood'.[336] In a catalogue of texts and authors discussed by W. Lambert, Adapa appears as a mediator figure who passes on knowledge he received from the gods.[337] In this text, he is known as ᵐ*ú-ma-an-na a-da-pa*, the cuneiform equivalent to Greek Oannes, culture bearer at the time of the first king of history.[338]

The reference to elders appears to be an attempt to find a link to the primal beginnings, to which wisdom belongs, and where it was originally made available. Eliphaz seems to imply that he is more connected with the ancient traditions than is Job. The only way Job's knowledge could exceed that of Eliphaz would be if Job indeed listens in the divine council — an image commonly used in the ideology of prophecy, and an image the author of the book of Job finds adumbrated in the conception of the primal human.

Human Conflict with the Divine (11–13)

Has God spoken a secret? (11)

The MT of v. 11 is difficult, both lexically and grammatically. The text and translation of NRSV is as follows:

הַמְעַט מִמְּךָ תַּנְחֻמוֹת אֵל
וְדָבָר לָאַט עִמָּךְ

> Are the consolations of God too small for you,
> that he should speak a secret with you?

What is the meaning of *tanḥūmôt* 'consolations' here, and how does it fit into Eliphaz's argument?[339] Gordis and Pope have understood

---

[336]W. G. Lambert, "A Catalogue of Texts and Authors," *JCS* 16 (1962): 67, 73–74. Note that the same is said about Gilgamesh. See also the Excursus on Adapa above, at the end of chapter 2.

[337]Ibid., 73. Lambert demonstrated this on the basis of the wording of the phrase ascribing authorship to Adapa. Similar wording is used of the revelation of the Erra Epic to its "author," Kabti-ilāni-Marduk.

[338]Lambert points out that the personal name determinative appears on the first element of the double name; the second then is an epithet, 'wise'. For discussion and other occurrences, see "A Catalogue of Texts and Authors," 74.

[339]See BDB, 637c. The word most likely derives from *nḥm* 'to be sorry, console'.

God's consolations to refer to the doctrine of rewards and punishments: God's comfort is embodied in Eliphaz's reiteration that no suffering is unmerited. Job, on the other hand, is not consoled, but feels misrepresented.[340] Similarly, Dhorme understood consolations to refer to the "inspired" comfort of the three friends, which Job eschewed as 'miserable' in 16:2b,[341] and Driver and Gray saw a reference to Eliphaz's experience of revelation in 4:12.[342] N. H. Tur-Sinai took 'consolations' as a reference to the "god" experienced by Job in his dream-vision and disbelieved by his friends, thus yielding the sense: 'According to what you tell us the words of the divine apparition held too little comfort for you...'.[343] The word *tanḥûmôt* appears in 21:2, but is applied to the three friends: 'Listen carefully to my words that this might be your *tanḥûmôt*'. The context of Job 21:2 seems to suggest the term is used in the sense 'explanation'.

The grammatical construction of the second colon of v. 11 is not immediately clear. Traditionally, the Hebrew of v. 11 has been rendered according to the principle of synonymous parallelism, whereby the second colon expresses the first in different words. But given the nature of the idiom *hamᶜaṭ . . .wə-* (second colon) as it is found elsewhere, there is greater precedent to understand, 'Is *x* not enough that you must have *y* as well?'.[344] The idiom requires completion, and one

---

[340]Gordis (*The Book of Job*, 161) bases his interpretation on the apparent reference to this doctrine in vv. 14–16 and 25:4. He finds significant the "legal sense" imparted to the phrase by Job in 9:2. Pope, *Job*, 115.

[341]Dhorme, *A Commentary on the Book of Job*, 212.

[342]Driver and Gray, *A Critical and Exegetical Commentary on the Book of Job*, 135.

[343]"Throughout this chapter Eliphaz refers to some specific words of one particular "god" only: the words of the divine being whose voice Job claims to have heard in his dream vision." *The Book of Job: A New Commentary* (Jerusalem: Kiryath Sepher, 1957), 249.

[344]Cf. Gen 30:15; Num 16:9,10; Josh 22:17–18. Perhaps most instructive in this case is the very similar occurrence in Isa 7:13 in which the sense is evident. 'Is it too little for you to weary mortals, that you weary my God also?' Here, the prophet Isaiah, in posing his rhetorical question, sets up a foil for his primary assertion: 'You are wearying my God'. Again, the sense is "you are not satisfied with *x*, you must have *y*," *y* being a logical extension or stretching of *x*. Tur-Sinai (*The Book of Job*, 249) repointed *hmᶜṭ* as a Piel verb (with prefixed interrogative), based on the (repointed)

might expect it in the second colon of Job 15:11.[345] A reasonable solution for the second colon is to repoint *dābār* as a Piel verb *dibber* and to read *lʾṭ* as *lāʾṭ* 'secret', with the resulting translation: 'that he should speak a secret with you'.[346] Gen 31:29 provides an example of the use of the verb *dbr* with the preposition *ʿim*: 'Take heed that you *speak to* Jacob neither good nor bad'.[347]

Given the grammatical construction and the context, I find it most reasonable to render the whole: 'Are the consolations of God (that is, the commonly understood principles of how the universe works) not enough for you that he should speak a secret with you?'. Not only is this more sensitive to the Hebrew idiom as reflected elsewhere, but it is more consonant with the context presented in vv. 7 and 8. There is additional support for this reading in the words of Zophar in 11:6, where a different word appears for secret (*taʿălūmôt*):

5 וְאוּלָם מִי־יִתֵּן אֱלוֹהַּ דַּבֵּר
וְיִפְתַּח שְׂפָתָיו עִמָּךְ׃
6 וְיַגֶּד־לְךָ ׀ תַּעֲלֻמוֹת חָכְמָה

5 But O that God would *speak*,
and open his lips *to you*,
6 and that he would tell you the *secrets of wisdom*.

---

parallel verb in the second colon. He suggested the sense, 'Has [the god] given little?', with *tanḥūmôt* as the direct object. The only other instance of the verbal root in the Piel (Eccl 12:3) is intransitive, however; although not impossible, this proposal finds little support.

[345]For *min-* following an adjective as 'too much, too little', see GKC §133c.

[346]Cf. 1 Sam 18:22; see BDB, 532. AV interpreted *lʾṭ* as 'secret'; Driver and Gray considered this interpretation of AV and RV (margin) "erroneous" although they failed to state why. Many commentators have understood *lʾṭ (ləʾaṭ)* to be the noun *ʾaṭ* with a prefixed preposition *le-* (cf. BDB, 31d B). Tur-Sinai has interpreted this as coming from the root *lwṭ* 'to conceal' with a "redundant ʾ" giving a reading, 'and concealed anything from you?'. (249). Although I believe he is correct in his understanding of the root, his translation is unlikely, given the prepositional phrase *ʿimmāk*. What is more, it goes completely against the grain of the speaker's preceding comments. There is, however, clear grammatical precedent for the reading suggested here that is in better accord with the context.

[347]Dhorme (*A Commentary on the Book of Job*, 212) accepts *dbr* as pointed by MT and appeals to other instances in which the noun is followed by the preposition *ʾim*.

What follows in vv. 7 and 8 shows Zophar's skepticism regarding the likelihood of seeing his wish fulfilled. Such is the sentiment revealed in the generally sarcastic tone of Eliphaz in our passage. As Eliphaz sees it, Job has expressed the sentiment that he possesses wisdom greater than that embodied in the generally-known, broadly-defined precepts considered available to all humans: a wisdom like that of the primal human. Further indication of this perceived sentiment of Job may be found in the somewhat curious words of 29:4, where Job reminisces about the past as a time he describes as *bəsôd ʾĕlôᵃh ʿălê ʾohŏlî*, 'when the *sôd* of God was upon my tent'.[348]

Heart, eyes, spirit, and speech: conflict with God? (12–13)

12 מַה־יִּקָּחֲךָ לִבֶּךָ
וּמַה־יִּרְזְמוּן עֵינֶיךָ
13 כִּי־תָשִׁיב אֶל־אֵל רוּחֶךָ
וְהֹצֵאתָ מִפִּיךָ מִלִּין

12 What has your heart instructed you?
How your eyes flash!
13 For you turn your spirit against El,
and you send forth words from your mouth!

We have seen that the primal human seems to find himself in a state of conflict with God. A similar idea seems present in the words of Eliphaz here. In v. 12 Eliphaz makes a statement that appears clear, yet there are questions about his precise intent. The versions and most critics understand this verse as an allusion to Job's pride and anger.

Heart and eyes (12)

Most authorities have rendered the particle *ma(h)* of v. 12 'why?' (a few read 'what?'; see below). It makes better sense to take

---

[348]NRSV understands *sôd* as familiar converse, and translates, 'when the friendship of God was upon my tent', which may in fact be the safest translation (cf. Ps 25:14 and Prov 3:32, neither of which requires a meaning of 'secret'). Nonetheless, one may not easily dismiss the phrase *sôd ʾĕlôᵃh* here as totally unrelated to the same phrase in 15:8.

the particle in its asseverative capacity, however,[349] although there is considerable ambiguity surrounding the use of the particle, and in each case context must be a guide.

As for the meaning of the phrase *yiqqāḥăkā libbekā*, there are no other instances in the Bible of the verb *lqḥ* with *lēb* as the subject. There have been many attempts to explain this phrase. Dhorme understood it to mean Job is "carried away by passion beyond the bounds of good sense." This he based on Prov 6:25, which he felt reflected the same sense.[350] Tur-Sinai rejected this interpretation and instead offered, 'What hath taken (from) thee thy reason?', construing *lēb* as a direct object (based on Hos 4:11) and the pronominal suffix *-k* as an indirect object with ablative force (based on Judg 17:2, *luqqaḥ-lāk* = 'taken *from* you').[351] Appealing to the non-standard use of *lk* (if taken as 'from you') renders this explanation improbable.[352] There appears to be no basis for interpreting the pronominal suffix attached to the verb here as anything other than a direct accusative.[353] Gordis' translation, 'Why does your passion inflame you?', is based on the phrase *ʾēš mitlaqqaḥat* in Exod 9:24 and Ezek 1:4. From a grammatical standpoint, however, this use is quite dissimilar.[354] Furthermore, since *mitlaqqaḥat* qualifies *ʾēš* and nowhere implies a connection with fire independently, there are no grounds for ascribing to the root *lqḥ* an intrinsic association with fire.

A reasonable possibility comes out of the sense which led to the semantic creation of the noun *leqaḥ*, commonly translated 'teaching' (cf. Deut 32:2, also Job 11:4), but which probably expresses literally

---

[349]See GKC §148; BDB, 553. According to the traditional rendering, Eliphaz asks a question and then answers it, as if he were saying "why *x*? because y!" This takes the *kî* of v 13 causally. The NRSV translation 'so that' expresses the sense of result. This, however, requires an unorthodox understanding of the conjunction *kî*.

[350]Dhorme, *Job*, 212.

[351]*The Book of Job*, 249.

[352]In Judg 17:2, the phrase may still simply reflect 'with respect to you' or 'belonging to you'.

[353]GKC §57.

[354]The construction to which he appeals is a Hithpael participle with no direct object.

'what is received'.[355] The verb appears in the Pual. In a hypothetical Piel it might then read, 'What has your heart *made* you receive?', that is, 'What are you convinced of?'. An idiomatic rendering would be, 'What has your heart instructed you?'. If the Qal verb could carry this more transitive meaning, then such an understanding is supported by v. 9, of which this might be taken as a restatement; this also explains the Tg reading, 'What does your thought teach you?'. Within the context of the primal human, this also echoes what we will see in chapter 5 in the statements against the *ʾādām* figure in Ezek 28:2 (cf. 28:6): *tittēn libbəkā kəlēb ʾĕlōhîm* 'You have made your mind like the mind of God'.

Any reasonable reading of *yirzəmûn* strongly suggests it to be something either "against" God, or that God would not like. This is clear whether one follows the Greek reading *epēnegkan* and emends *yirzəmûn* to a verb from the root *rwm*, 'to be high, haughty',[356] or accepts that the correct root is *rmz* (assuming a metathesis) and Driver and Gray's observation linking this phrase to Sir 27:22, 'winking' the eye in an evil sense.[357] Tur-Sinai interpreted the root according to Arabic *razama*, 'to dwindle away, to become weak' and argued that the pair 'heart and eyes' refers to the intellect and the senses. Thus, he gave the translation, 'Why have thine eyes weakened?'; that is, 'Why have you lost your perceptive faculties...have you become deranged?'.[358] It seems clear regardless that Eliphaz accuses Job of some manner of affront to God.

---

[355]BDB, 544.

[356]BDB, 931. Cf. Prov 6:17 and 30:13, both of which express negative states or actions.

[357]The metathesis is reflected in Tg, Syr, and some medieval Hebrew manuscripts. Driver and Gray observed that Aramaic *rmz* corresponds to Greek *dianeuō* (*A Critical and Exegetical Commentary on the Book of Job*, 96–97; followed by Dhorme [*A Commentary on the Book of Job*, 213]). Gordis also accepted the metathesis *rmz* 'to wink, flash' (*The Book of Job*, 161).

[358]*The Book of Job*, 250. Followed by Pope (*Job*, 116). The pair 'heart and eyes' is used to express a wide range of meanings and is in no way restricted to intellect and senses (at least not in the way Tur-Sinai has taken it). In some cases it is an unequivocal reference to emotions (1 Sam 2:33; Lam 2:11, cf. 5:17, Deut. 28:65, Ps. 19:8). The pair cannot, therefore, be considered significant evidence in elucidating this verse.

Confrontation with El (13)

To consider these verses as belonging to the framework of the first human reveals the figure to be embroiled in controversy. In the first colon of v. 13, *rûḥekā* can be understood either as the subject or the object of the verb *tāšîb*. In the vast majority of cases, the Hiphil of the root *šwb* with an accusative of object followed by the preposition *ᵓel* expresses the meaning to return a person or thing to a person or place, and critics have traditionally rendered this verse, 'You have turned your spirit against God'. The best witnesses for the translation of the Hiphil of the verb *šwb* with *ᵓel* as 'to turn against' are Amos 1:8; Isa 1:25; Zech 13:7; and Ps 81:15. They all refer to the same idiom, however: 'to turn one's hand against'; and in each case it refers to punishment by Yahweh.

Pope cited Prov 26:32; 25:28; and 29:11 as examples of the use of spirit/wind in the sense of anger. These examples are distant from the idiom expressed here, however. Likewise Driver and Gray cited Judg 8:3.[359] Tur-Sinai offered a creative solution by interpreting שׁוב with אל as ascribing one thing to another, and rendered 'that thou ascribest thy vapouring to a god'. He cited as evidence Akkadian *turru,* meaning 'to turn, convey' as well as 'to bring back'.[360] The meaning is that Job has "put [his] own futile word into the mouth of the god whose voice [he] claim[s] to have heard." He rendered the second colon as an introduction to what he understood as a quotation in vv. 14–16: 'and utterest from thy mouth the words...'. This last suggestion is highly unlikely, given the fact that the words of vv. 14–16 constitute what is properly understood as a common saying; it appears twice elsewhere in Job, slightly modified, and is spoken by Eliphaz and Bildad, respectively. I will address the content of this saying below.

Some attention must be given to the words alleged to issue from Job's mouth. Eliphaz does not elaborate beyond his simple reference to 'words'. Throughout the book of Job, the hero maintains his inno-

---

[359]*A Critical and Exegetical Commentary on the Book of Job,* 135. See also Gordis, *The Book of Job,* 161.

[360]*The Book of Job,* 250.

cence. A major theme of the comforters is that his misfortune is a result of his state of uncleanness or sin. It may be safely inferred that the 'words' of which Eliphaz speaks are no less than Job's persistent claim of righteousness and God's unfair treatment of one who is inculpable (9:30,31; 11:4; 16:17; 33:9). This is precisely the thing upon which he relentlessly insists, along with his implied wisdom or knowledge (cf. 11:4a), that makes his charge against God legitimate — so much so that he believes he can win his case if decided by an impartial arbiter (9:33; 16:19). The essence of the 'words' is captured within the context of this passage. Speaking of the calamity suffered by the wicked, he gives as a reason in 15:25: 'because he has extended his hand against God (*kî-nāṭâ ʾel-ʾēl yādô*) and behaved proudly toward the Almighty (*ʾel šadday yitgabbār*)'.[361]

Perhaps more than anything, Job is guilty of a claim to access to truth beyond normal human limits. Speech here does not have to comprise a direct challenge to the dominion of God (cf. Isaiah 14), but Job's alleged wisdom does allow him to impugn the integrity of God. It is a fundamental component of Israelite belief that Yahweh, as God, must be above reproach. This belief drives the corpus of literature designated Deuteronomistic, for instance. It is an inescapable conclusion, then, that possession of the secrets of the cosmos can only put a human being on a collision course with the creator, or present a perpetual situation of potential conflict, precisely because humanity is not above reproach. The potential for the abuse of power is ever present. The primal human traditions are about the abuse of wisdom, which is in effect, an abuse of power, which in turn highlights humanity's free will.

There are clearly some translational difficulties in this section; nonetheless, the versions all agree that Eliphaz has alleged some type of confrontation between Job and God, and I have argued that that confrontation is a part of Eliphaz's first human allusion.

---

[361]Cf. Job 36:9.

The Uncleanness of Humankind (14–16)

It is reasonable to understand these next three verses as an elaboration of the preceding verses. The tenor of the comments about Job, the first human, the council of El, and wisdom in vv. 7 and 8 resonates in the statements concerning Job's attitude toward God in vv. 11–13. Each section explores the relationship between Job and God; or better, perhaps, Job and deity. What is at stake is Job's right to challenge the actions of God; and the validity of that right is predicated upon what is, in essence, Job's claim to wisdom — a higher knowledge that puts him on a par with God as a judge of the morality of actions. In the first section Job is in the divine council; in the second, he is in confrontation with God; in the third, the relationship is defined — resolved — in favor of God.

This last section must be understood alongside 4:17–19, a passage in which a similar statement was spoken to Eliphaz by a shadowy being in a dream-vision, and 25:4–6 in which the saying of 15:14 is repeated almost verbatim. Read alongside the preceding verses, v. 15 seems to assert that even if Job were the first human (or any privileged being), he would still be considered unclean and not worthy of divine confidence. The notion of uncleanness in this first-human context is immediately reminiscent of the same occurring in other like passages, most notably the oracles of Ezekiel 28. Job 15 and the oracles in Ezekiel 28 are similar in that they both feature a superior type of human and an identifiable human analogue: in Job 15, the superior being is the first human, to whom Job is compared and contrasted; in Ezekiel 28, the prince and king of Tyre are assimilated by metaphor into the primal human paradigm. In both Job and Ezekiel there is an apparent correlation between wisdom and the concept of uncleanness. We have seen that uncleanness plays a role in the Adapa myth as well. I will return to this correlation shortly; first, however, it remains to examine the parallel passages in Job 4:17–19 and 25:4–6.

Placed alongside one another, the three passages reveal some differences. The variations may be accounted for in a number of ways. Equally interesting are what they share in common:

4:17–19	15:14–16	25:4–6
4:17a Can a mortal be *righteous (yiṣdaq)* before God?	15:14a What is a man that he can be *pure (yizkeh)*?	25:4a How then can a man be *righteous (yiṣdaq)* before God?
4:17b Can a man be *pure (yiṭhar)* before his maker?	15:14b Or he that is born of a woman that he can be *righteous (yiṣdaq)*	25:4b How can he who is born of a woman be *pure (yizkeh)*?
4:18a Even in his *servants* he puts *no trust (lōʾ yaʾămîn)*	5:15a He *trusts not (lōʾ yaʾămîn)* even in his *holy ones*	25:5a Behold, even the *moon* is *not bright (lōʾ yaʾăhîl)*
4:18b and his *messengers* he charges with error	5:15b And the *heavens* are *not clean* in his sight *(lōʾ zakkû bəbəʿênāyw)*	25:5b and the *stars* are *not clean* in his sight *(lōʾ zakkû bə-ʿênāyw)*
4:19a How much more *(ʾap)* those who dwell in houses of clay	15:16a How much less *(ʾap kî)* one who is abominable and corrupt	25:6a How much less *(ʾap kî)* a man, who is a maggot
4:19b whose foundation is in the dust, who are crushed be-fore the moth	15:16b a man who drinks iniquity like water	25:6b and a human who is a worm

The three passages present what is in essence the same saying. Each is composed of three bicola, with the possible exception of 4:17–19, in which v. 19 may be construed a tricolon.

Purity and righteousness (14)

The sense of v. 14 is that no human is either pure or righteous. It is a summary introduction to what follows in vv. 15 and 16.

מַה־אֱנוֹשׁ כִּי־יִזְכֶּה
וְכִי־יִצְדַּק יְלוּד אִשָּׁה

What is a man that he can be pure?
Or that he can be righteous, he that is born of a woman?

In each of the three passages, the verb *yiṣdaq* appears and is paralleled by a verb conveying the idea of purity. *yiṭhar* in chapter 4, and *yizkeh* in chapters 15 and 25; *yiṣdaq* is the second element of the parallel relationship in chapter 15 (*yizkeh-yiṣdaq*), whereas it is the first element in chapters 4 and 25 (*yiṣdaq-yiṭhar* or *-yizkeh*). The verb *yizkeh* (from the root *zkh*) 'to be clear, clean, pure' is always used within the context of morality in the Hebrew Bible. It appears twice in Job (both in the G) meaning to be clean, pure in the sight of God (both instances occurring in the passages now under consideration).[362]

'No trust in his holy ones' (15)

The second bicolon of the saying in 4:18; 15:15; and 25:5 is best explained, I believe, on the basis of its manifestation in chapter 15: *hēn biqdōšāw lōʾ yaʾămîn wəšāmayim lōʾ-zakkû bəʿênāyw*, 'God trusts not even his holy ones, And the heavens are not clean in his sight.' The first colon of 15:15 shares the phrase *lōʾ yaʾămîn* with 4:18; in both cases, it applies to God's attitude toward the supra–human or angelic realm (servants and holy ones). Here, 25:5 differs by using the phrase *lōʾ yaʾăhîl* and applying it to the moon. The second colon of 15:15 shares the phrase *lōʾ zakkû bəʿênāyw* with that of 25:5; in both cases, it applies to God's attitude toward the heavenly host in their concrete, visible form: in 15:15b it is the heavens and in 25:5b it is the stars.[363] The corresponding colon of 4:18b differs by using the phrase *yāśîm tohŏlâ bə-* 'he ascribes error', and applies it to God's attitude toward angels. In this portion of the saying, chapter 4 considers "personal" beings, whereas chapter 25 considers what appear to be impersonal phenomena. The second bicolon of the saying as preserved in chapter 15 (15:15a,b) may be considered an interpretive linchpin, tying the three passages together and casting light on the other two: it shares a verbatim phrase both with 4:18 *(lōʾ yaʾămîn)* and with 25:5 *(lōʾ zakkû bəʿênāyw)*; and it combines references to both personal beings and impersonal phenomena. These latter are only apparently impersonal,

---

[362]BDB, 269. The verb has cognates in Akkadian *zakû* 'to be clear' (referring to elements such as water or sky) and 'to be clean or pure' (referring to clothes, persons, or metals) and Aramaic *dəkāʾ* meaning 'to be pure, innocent, upright'.

[363]The verb *zakkû* is traditionally traced to the root *zkk*, which is generally translated 'to be bright, clean, pure' (BDB, 269). It is recognizably related to the root *zkh*, which we have already seen.

however, and should be considered akin to other similar references to celestial phenomena (e.g., invoked as witnesses).[364] There is ample evidence of these phenomena signifying deities, Deut 4:19, for instance.[365] In Job 38:7 the morning stars (*kôkəbê bōqer*) are placed in a parallel relationship with the sons of God, *bənê ʾĕlôhîm*.

The phrase *hēn biqdošāyw / baʿăbadāyw lōʾ yaʾămîn*, 'In his holy ones/servants he puts no trust' (15:15a and 4:18a), is curious, and important in discerning meaning in this passage. There is considerable ambiguity surrounding the precise meaning of the root *ʾmn*.[366] It appears several times in the book of Job — some instances more easily understood than others. Nonetheless, generally speaking, we can be somewhat more sure of deriving the meaning of a word or phrase within the context of a single book. I believe the use of the verb in Job 15:15 may be best explained on the basis of 39:9–12, which the NRSV translates as follows:

9 הֲיֹאבֶה רֵים עָבְדֶךָ אִם־יָלִין עַל־אֲבוּסֶךָ
10 הֲתִקְשָׁר־רֵים בְּתֶלֶם עֲבֹתוֹ אִם־יְשַׂדֵּד עֲמָקִים אַחֲרֶיךָ
11 הֲתִבְטַח־בּוֹ כִּי־רַב כֹּחוֹ וְתַעֲזֹב אֵלָיו יְגִיעֶךָ
12 הֲתַאֲמִין בּוֹ כִּי־יָשׁוּב [יָשִׁיב] זַרְעֶךָ וְגָרְנְךָ יֶאֱסֹף

9 Is the wild ox willing to serve you?
Will he spend the night at your crib?
10 Can you bind him in the furrow with ropes,
or will he harrow the valleys after you?
11 Will you depend on him because his strength is great
that you can leave to him your labor?
12 Do you have faith in him *(taʾămîn bô)* that he will return,
and bring your grain to your threshing floor?

---

[364] A. Negoită and H. Ringgren (*TDOT* 4.63) take chapter 25 to be a purely metaphorical application of the literal meaning of the root *zkh*; that is, *yizkeh* (and *yaʾăhîl*) refer simply to the "clear and bright illumination of the heavenly bodies." I would suggest on the basis of the other passages, however, a deliberate double entendre.

[365] That this interpretation applies to Job 25:5 may be seen in the verses which precede the saying, which set the passage within the context of Yahweh's relationship with the heavenly host (vv. 2,3).

[366] It appears in the Qal, Niphal, and Hiphil, and there is debate over whether these recorded instances in the Hebrew Bible represent more than one root. Our interest, of course, is with the Hiphil use of the root.

The point of this passage is that the ox is a creature of its own will. *heʾĕmîn bə-* is used to denote belief or faith, in the sense of 'trust'. Can one trust a creature of free will?

Another occurrence of the phrase *lōʾ yaʾămîn bə-* in Job germane to this discussion is 24:22. Here the text reads *lōʾ yaʾămîn baḥayyîn,* which NRSV translates, 'they despair of life'. The same phrase appears in Deut 29:66 in a similar context, translated by NRSV, 'have no assurance of life'. This passage refers to a fear that life will not continue — not a knowledge of one's immanent demise so much as a lack of knowledge about life's future.

What does it mean that the heavens are not clean? It is unlikely that the uncleanness of the heavens reflects an understanding of a creation brought about with inherent flaws. In fact, in view of the notion of a "good" creation presented in the opening chapter of Genesis, this statement comes as quite a shock. Exod 24:10 reveals the conception of a heaven associated with or even defined by the notion of purity: 'And they beheld the God of Israel; beneath his feet was something like a work of sapphire brick, like the essence of the heavens with respect to purity' (*kəʿeṣem haššāmayim lāṭōhar*). Here, the idea of the purity or clarity of the heavens is such that 'heavens' can serve appropriately in a simile describing this quality in something else. This verse appears to tell us "the heavens are clean" — in essence, the very opposite of what Eliphaz seems to be expressing in Job 15:15. Exod 24:10 and Job 15:15 are best taken as referring to two different ways of understanding the term "heavens." Exod 24:10 speaks of heavens with regard to simple physical appearance; Job 15:15 speaks of heavens as a reference to spiritual beings, or deities. The context of Exod 24:10 is a description of the physical appearance of the thing upon which the God of Israel stood. The context of Job 15:15 on the other hand has nothing to do with the simple physical qualities any type of object; rather, its concern is within the realm of morality.

It is very likely that Job 15:15 reflects the notion of creation within the context of a cosmic struggle — an image well known in the Mesopotamian materials. The presence of this motif has been detected in the Hebrew Bible as well, although the extent to which it is present is debated. Recently, J. Levenson has demonstrated the prominence of the cosmogonic struggle in Israelite literature, making clear that the cosmogonic battle won by Yahweh at creation amounted to limits being placed upon chaotic forces that perpetually threaten to encroach upon the order he established *in illo tempore* and maintains in history.[367]

The idea that God puts no trust in his *holy ones* is striking and is not far from the category of oxymoron. It is particularly striking given that there is precedent for the more neutral (with respect to morality) designation 'servant' in the corresponding verse of the parallel passage in chapter 4. Likewise, 'angels' would be a less dissonant alternative. The context of Eliphaz's argument lends strong support to the conclusion that the reading of the consonantal MT 'his holy one' was not the result of a scribal error (*bqdšyw* → *bqdšw*), but rather reflects the original intent of the popular saying as it was modified to fit the argument in this particular context, perhaps the result of a scribal tradition that understood it as a reference to the primal human.

No trust in a human (16)

It is in v. 16 that Eliphaz brings home his point with Job.

אַף כִּֽי־נִתְעָב וְנֶאֱלָח
אִישׁ־שֹׁתֶה כַמַּיִם עַוְלָה

How much less one who is abominable and corrupt,
A man who drinks iniquity like water.

In Job 15:16 humanity is referred to as *nitᶜāb* and *neʾĕlāḥ*. The participle *nitᶜāb* comes from the verbal root *tᶜb*, a denominative created from the noun *tôᶜēbâ* 'abomination'. In the Niphal the meaning is 'to be abhorred', and in the Hebrew Bible it is employed both in a ritual sense and ethically. The word *neʾĕlāḥ*, from the root *ʾlḥ* appears only in the Niphal, and seems to bear the distinct sense of moral corruption.[368]

---

[367]J. Levenson, *Creation and the Persistence of Evil* (San Francisco: Harper and Row, 1988), see especially part 1 (*passim*). Levensen explains the eventual demythologization of the primordial chaotic enemy (whereby it becomes a created being) in chapter 5 (53–65).

[368]The meaning of אף כי is evident by the context. The phrase is used to mean

The purpose of the saying is not simply to excoriate and degrade humanity. The sayings in chapters 4 and 25 set up a comparison between God and humanity: how can humanity be pure before God, or be more pure than God? Chapter 15 does the same. It is the purity of God versus the impurity of humanity. And it is Job's impurity that the comforters argue is the source of his wretched state.

Within this equation are the heavens, which are also reckoned as unclean. Verse 15 is by no means a simple affirmation that "everything is bad"; rather it sets up God as the only being with perfect purity and integrity. What is interesting here is that the first human, who is primordial, lies somewhere between God and human on what might be called the deity — humanity continuum. He is not born of woman, but neither can he be reckoned as God. In chapter 15 he is reckoned with the unclean and untrustworthy.

Ultimately, I believe this passage, along with the invocation of the primal human, is a subtle reflection on the abuse of power. In this passage, it is wisdom which is the power subject to abuse. In the book of Job God alone remains above the fray. Humankind is not a party to the secrets of the cosmos, for in its hands they are subject to abuse. The primal human is counted among the divine beings not worthy of trust. The primal human is striking here because he represents the linkage between humanity and God.

A text I find to be quite germane to our discussion at this point is Psalm 8. Discerning the meaning of this text has vexed scholarship for quite some time. It has been interpreted as a hymn in praise of creation as well as a paean to the king. What I would like to point out is that this text bears significant similarities with the saying which is the present object of study. The saying as it is found in Job 25:4–6 is the manifestation which most invites comparison. In this psalm are found three similar elements: a reference to heavens, moon, and stars; a comparison drawn concerning humanity; the comparison is accomplished using the *ʾĕnôš / ben ʾādām* formula. I am inclined to argue that the reference

---

'furthermore, indeed', '(is it) that?' (in a question). When used in conjunction with a preceding sentence (which is often introduced by *hēn* or *hinnēh*), it carries the force of 'how much more'. Prov 11:31 similarly expresses an idea not unlike that found here, also beginning the first clause with the particle הן.

to celestial phenomena is of the same nature as the passage in Job 25. We find a rather similar use in Ps 148:2–4. I might also conjecture that the author of Psalm 8 was familiar with our saying, and included a modified variant of it in his psalm.[369]

Summary of Primal Human Topoi in Job

In the world view of the writer of Job and his audience the first human is an exalted being. This nature is expressed in the speech of Eliphaz, and alluded to in the words of the theophany. He is numbered among the sons of God, which, as we have seen (chapter 1), is an idea expressed in the Sethite genealogy of Genesis. Despite his obvious connection to humanity, he is considered something significantly different. He is a foil to the unenlightened mortal.

The location of the primal human

The speech of Eliphaz presents the primal human in the council (*sôd*) of God. The *sôd* of God — an assembly of deity — was a *place for divinity*, and not humanity, and it is from this point of tension that Eliphaz begins to make his point. We saw that the language chosen by Eliphaz had a clear basis in Israelite society. The language was precisely that used by those involved in the prophetic realm. It was in every respect unusual for a human being to enter the company of deity, and it was this special aspect that gave prophecy its legitimation. Job is by no means a "prophetic" book; there is no apparent pro-prophecy agenda. Yet for his general audience Eliphaz locates the basis of this prophetic imagery in the myth of the primal human.

The wisdom of the primal human

According to Eliphaz, the wisdom of the primal human came as a result of his presence within the council of God, and the fact that he 'listened'. The knowledge at issue belongs to the gods, and not to

---

[369]To be sure, the man/son of man formula did not belong to a single saying; it is found independently in a number of contexts (e.g., Job 7:17–18; Ps. 144:3,4).

common humanity. We are reminded again of the woman who praised David's divine knowledge, likening him to the messenger of God who discerns (literally 'hears') good and evil.[370] The wisdom of which Eliphaz speaks is higher knowledge, that which would explain Job's situation, but goes far beyond. The book of Job clearly reflects an understanding of the primal human as a paragon of wisdom.

Creation and the primal human

We have seen how this higher knowledge is connected with the primordium, and is closely associated with creation. The primal human is no mere usurper; rather, he was present at the creation and by virtue of that fact possessed wisdom in its most intimate details. The divine speeches in chapters 38–41 make clear that the secrets of the universe lie within the primordium, the epoch of creation. As one who 'was born then', he knew the deepest and most esoteric of knowledge.

Conflict and the primal human

Despite the positive attributes the primal human may possess, he is nonetheless presented within a curiously tension-filled context. According to the text, the issue of "purity" or "cleanness" plays a significant role in creating this tension. Within this context of conflict, human is pitted against God, with the divine created beings occupying a position in the middle. It is in this middle position that we find the primal human; and it is here that we learn that even he 'is not pure before God'.

---

[370] 2 Sam 14:17, which we have explained in chapter 2 as a reference to comprehensive divine knowledge; see also 2 Sam 14:20.

Part III

Vestigial Allusions

The Sublimation of Primal Human Imagery

5. The Prince of Tyre: Adam and not God (Ezekiel 28:1–10)

EZEKIEL 28:1–10, TOO, IS A PART of the primal human tradition, but several particulars bear noting. Unlike the narratives in Genesis 1 and 2 (examined in Part I), it does not present the tradition by relating information about the protagonist in a direct fashion. Like the oracle in Ezek 28:11–19 and the passage in Job 15 (examined in Part II) it applies the imagery analogically to another figure. However, the ideas have been further sublimated — so much so that what remains appears to be quite unrelated to the traditions of Genesis, Ezekiel, and Job. In spite of these apparent differences, however, there are unmistakable elements that reveal the essential background of the imagery — and that background is indeed that of the primal human.

The Text: Ezekiel 28:1–10

1 וַיְהִ֥י דְבַר־יְהוָ֖ה אֵלַ֥י לֵאמֹֽר׃
2 בֶּן־אָדָ֡ם אֱמֹר֩ לִנְגִ֨יד צֹ֜ר כֹּֽה־אָמַ֣ר ׀ אֲדֹנָ֣י יְהוִ֗ה
יַ֣עַן גָּבַ֤הּ לִבְּךָ֙ וַתֹּ֙אמֶר֙ אֵ֣ל אָ֔נִי מוֹשַׁ֧ב אֱלֹהִ֛ים יָשַׁ֖בְתִּי בְּלֵ֣ב יַמִּ֑ים
וְאַתָּ֤ה אָדָם֙ וְֽלֹא־אֵ֔ל וַתִּתֵּ֥ן לִבְּךָ֖ כְּלֵ֥ב אֱלֹהִֽים׃
3 הִנֵּ֥ה חָכָ֛ם אַתָּ֖ה מִדָּנִאֵ֑ל כָּל־סָת֖וּם לֹ֥א עֲמָמֽוּךָ
4 בְּחָכְמָֽתְךָ֙ וּבִתְבוּנָ֣תְךָ֔ עָשִׂ֥יתָ לְּךָ֖ חָ֑יִל וַתַּ֤עַשׂ זָהָב֙ וָכֶ֔סֶף בְּאוֹצְרוֹתֶֽיךָ׃
5 בְּרֹ֧ב חָכְמָתְךָ֛ בִּרְכֻלָּתְךָ֖ הִרְבִּ֣יתָ חֵילֶ֑ךָ וַיִּגְבַּ֥הּ לְבָבְךָ֖ בְּחֵילֶֽךָ׃
6 לָכֵ֕ן כֹּ֥ה אָמַ֖ר אֲדֹנָ֣י יְהוִ֑ה יַ֛עַן תִּתְּךָ֥ אֶת־לְבָבְךָ֖ כְּלֵ֥ב אֱלֹהִֽים׃
7 לָכֵ֗ן הִנְנִ֨י מֵבִ֤יא עָלֶ֙יךָ֙ זָרִ֔ים עָרִיצֵ֖י גּוֹיִ֑ם וְהֵרִ֤יקוּ חַרְבוֹתָם֙ עַל־יְפִ֣י
חָכְמָתֶ֔ךָ וְחִלְּל֖וּ יִפְעָתֶֽךָ׃
8 לַשַּׁ֖חַת יוֹרִד֑וּךָ וָמַ֛תָּה מְמוֹתֵ֥י חָלָ֖ל בְּלֵ֥ב יַמִּֽים׃
9 הֶאָמֹ֤ר תֹּאמַר֙ אֱלֹהִ֣ים אָ֔נִי לִפְנֵ֖י הֹרְגֶ֑ךָ וְאַתָּ֥ה אָדָ֛ם וְלֹא־אֵ֖ל בְּיַ֥ד
מְחַלְלֶֽיךָ׃
10 מוֹתֵ֧י עֲרֵלִ֛ים תָּמ֖וּת בְּיַד־זָרִ֑ים כִּ֚י אֲנִ֣י דִבַּ֔רְתִּי נְאֻ֖ם אֲדֹנָ֥י יְהוִֽה׃

1 The word of Yahweh came to me saying:
2 "Son of man, say to the prince of Tyre, 'Thus declares the lord Yahweh:
"Because your heart became exalted and you declared 'I am God, I sit

in the dwelling of the gods in the heart of the seas' —
yet you are man and not God, though you have made your mind like the mind of gods.

3 Indeed you are wiser than Daniel, no secret is hidden from you.

4 With your wisdom and with your understanding you made for yourself wealth, and you have made gold and silver for your treasuries.

5 With your great wisdom in your trading practices you increased your wealth, and your heart has become exalted in your wealth."

6 Therefore thus declares the lord Yahweh: "Because you made your mind like the mind of gods;

7 therefore, I am sending against you foreigners, tyrants of the nations; They will draw their swords against the beauty of your wisdom, and they will defile your splendor.

8 They will send you down to the pit, and you will die the death of those who are profaned in the heart of the seas.

9 Will you say 'I am God' before your killers, though you are human and not god in the hands of those who defile you?

10 You will die the death of the uncircumcised at the hand of foreigners, for I have spoken," declares the lord Yahweh.

The Relationship between Ezekiel 28:1–10 and 11–19

The question of the relationship between the two major oracles of this chapter has always played an important role in the scholarly debate.[371] I would like to suggest here that the two oracles feature the same mythic character: *ʾādām*, the ancient Israelite primal human. The first is a "thematic" version of the myth, if it in fact can be called a version at all. It amounts to the presentation of a single statement from the myth, which is then embellished with traditional imagery concerning Tyre. This apocopated form of the myth provides what is, in essence, a summary of what is perceived to be the point of the myth — at least for the prophet's purposes here. This, in part, explains the relationship between the two oracles: the second oracle is the more developed in its presentation of details of the original myth.

The two oracles were undoubtedly composed together. They are clearly presented as two separate oracles, but they nonetheless comprise a well-conceived unit. The first is presented as an oracle which predicts the downfall of Tyre. The second oracle is a "post-mortem"

---

[371] For a recent summary of scholarship and discussion on the relationship between 28:1–10 and 11–19 see H. R. Page, *The Myth of Cosmic Rebellion: a Study of Its Reflexes in Ugaritic and Biblical Literature* (Leiden; New York: E. J. Brill, 1996).

reflection after the city's demise. It is clear that vv. 1–10 and 11–19 are presented as separate oracles simply to draw attention to the prophetic concern regarding prediction and fulfillment. This arrangement is clouded by two differences: first, vv. 1–10 are addressed to the prince of Tyre, and vv. 11–19 to the king; second, the fact that the predictions of the first are not literally restated in the second. The conclusion may nevertheless be maintained on the basis of form-critical and grammatical observations. The first oracle is a standard *announcement of judgment* which contains the traditional elements of *reproach* (or *invective*) and *threat*. It is a prediction of the future: the threat (vv. 6–10) is expressed by means of a series of verbs in the imperfect. The second oracle is a *lamentation* and is presented largely in the form of an historical retrospect (in figurative terms). The announcement of punishment (vv. 16–19; corresponding to the threat of the first oracle) is expressed in a series of verbs in the *wāw* consecutive. The consequences of the sin of the king of Tyre are presented as an accomplished fact.[372]

Identifying Mythological Allusions in Ezekiel 28:1–10

In both oracles the tenor and vehicle of the metaphor are not easily distinguished. It is important to detect which elements belong to the myth complex as well as the nature of the interface between the primal human and the literal situation addressed by the prophet. Part of this task includes examining Ezekiel's words in the light of the book of Ezekiel as a whole, and other prophetic writings. In each oracle, Ezekiel presents statements from the mythic complex that the audience would immediately recognize. These statements are then set into a framework which both transforms it into a message against Tyre and elucidates the nature of the analogy between Tyre and the primal human. This is achieved through the combining of three elements: a) traditional prophetic rhetoric of a *general* nature (that is, imagery and statements which can apply to anyone); b) traditional prophetic rhetoric of a *specific* nature (that is, in this case, traditional statements concerning Tyre); and c) *mythological allusions* — statements from a

[372]Compare Ezekiel 26, 27. Zimmerli also recognizes the pattern of divine judgment oracle followed by lamentation in chapters 17 and 19; chapters 29–31 and 32 (*Ezekiel* 2.87).

commonly understood mythic complex. For convenience, I will refer to these as *general statements*, *specific statements*, and *mythological allusions*, respectively.

Certain elements are more readily discerned than others. *General statements* may be of a highly routinized nature and "proverbial" (or formulaic) and as such, they may appear verbatim in other passages. For this reason, they are obviously far more easily recognized. But they may also have parallels elsewhere which are an expression of the same idea in different language (this is much of the basis of form criticism). Still, the size of the prophetic corpus within the Hebrew Bible and the discipline of form criticism make these general statements the most easily discerned of the three elements. Certain ideas and phrases both recur repeatedly and are applied to many different subjects. *Specific statements* are slightly more elusive because the subjects to which they apply obviously comprise a fraction of the total number of possible subjects. In other words, some ideas and phrases are traditionally applied only to specific nations or figures and therefore occur less frequently in the corpus as a whole (as compared with general statements). *Mythological allusions* are the most difficult; they are analogical in nature and are often left unexplained. These must also be sorted out from tropes which are the writer's own creation.[373] Mythological allusions are identified by appeal to comparative materials, which is both logical and necessary, although often yielding somewhat tentative results. It is reasonable to expect some overlap in places: a phrase can reflect more than one element, and that overlap makes interpretation more difficult; but it also underscores the need to properly discern cases that are less ambiguous.

In the case of the first oracle (vv. 1–10), the fact that the text comprises these three elements and that they are not always clearly defined has resulted in a lack of scholarly consensus regarding its interpretation. Failure to distinguish properly between one element and another can produce all manner of fanciful interpretations; indeed how one identifies any given element can set the course for how the entire passage is viewed.

---

[373]One can always expect a degree of artistic license in presentation.

The Presence of Primal Human Themes

The wisdom of the primal human

The phrase *ʾēl ʾănî* 'I am God' here seems to signify *wisdom* insofar as it suggests the imagery presented in the words of God in Gen 3:22: 'The man *has become like one of us*, knowing good and evil'. In the oracle against the prince of Tyre, 'I am God' is explained by the phrase 'You have made your mind like the mind of gods' (*tittēn libbəkā kəlēb ʾĕlōhîm*) (v.2), which recurs in v. 6. Further explication of *ʾēl ʾănî* is set in between these two occurrences.

> Indeed you are wiser than Daniel,
> no secret is hidden from you.
> With your wisdom and with your understanding
> you made for yourself wealth
> and you have made gold and silver for your treasuries.
> With your great wisdom in your trading practices
> you increased your wealth,
> and your heart has become exalted in your wealth.

ʾēl in the phrase *ʾattâ ʾādām wəlōʾ ʾēl* here refers to God, that is, as opposed to a human. There is an obvious interpretive problem in the word *ʾēl* which has contributed to the confusion. It may be used as a divine proper name, the god El, who is known from Ugaritic and other sources; but not all of the proposed occurrences of this usage in the Bible have been universally accepted.[374] The phrase *ʾādām wəlōʾ ʾēl* was apparently in common parlance in ancient Israel (at least among the prophets), evidence for which may be found in the earlier Isaiah, where it appears verbatim, but in a different context. In Isa 31:3, the prophet says of the Egyptians in whom Judah sought refuge, 'The Egyptians are human and not God' (*ʾādām wəlōʾ ʾēl*). Here the phrase probably refers not just to Pharaoh's claim to divinity (which claim, incidentally is better documented than the Tyrian king's) but to the fundamental question of the human/divine distinction — a question that forms the basis of the Genesis account of the primal human.

---

[374]For a discussion of the range of uses, see Cross, "ēl," *TDOT* 1.253–60.

The location of the primal human: the dwelling of the gods

Most commentators take *môšāb ʾĕlôhîm yāšabtî* together with *bəlēb yammîm* as inseparable. The two phrases are indeed uttered as a single statement, but it is apparent that they have separate backgrounds. The phrase *bəlēb yammîm* is a traditional specific statement about Tyre. It appears four times regarding Tyre in chapter 27 (vv. 4,25–27). The statement in 27:4, 'Your borders are in the heart of the seas', explains the other three references in which Tyre is allegorized as a ship: it refers to the geographical location of the city.[375] The image of the sinking ship is an expression of the demise of the city Tyre in its littoral setting. Tyre was known in these terms internationally. In Papyrus Anastasis I, Tyre is called 'a city in the sea', and Ashurbanipal stated, 'I marched against Baal, King of Tyre who dwells in the midst of the sea' (*āšib qabal tâmtim*).[376] There is additional evidence to discourage the notion that the phrase might be used exclusively as code for the divine dwelling. The phrase *bəlēb yammîm* appears in the book of Jonah, where it stands for no less than the sea as a chaotic force, 'for you cast me into the deep, into the heart of the seas (*bəlēb yammîm*), and the flood was round about me'.[377]

bəlēb yammîm in 28:2 (and 8) is a *specific statement about Tyre,* which aids in the application of the mythic complex to Tyre. In chapter

---

[375]Note also 27:3 — Tyre dwells at the entrance to the sea.

[376]For discussion and references see M. Pope, *El in the Ugaritic Texts* (VTSup 2; Leiden: E. J. Brill, 1955), 98.

[377]M. Pope (*El in the Ugaritic Texts*, 98) understood the oracle to be based on the myth of the banishment of the Ugaritic god El from heaven to a watery abode. (See also U. Cassuto, *The Goddess Anath: Canaanite Epics of the Patriarchal Age* [trans. Israel Abrahams; Jerusalem: Magnes, 1971], 57.) Pope argued that the phrase *môšab ʾĕlōhîm...bəlēb yammîm* did not refer to the general dwelling of the gods; rather it signified "the specific abode of a god who dwells in watery environs." Following Cassuto, that god he thought to be none other than Ugaritic El. In order to maintain this interpretation, Pope (ibid.) considered the use of the phrase. 'in the heart of the seas', in v. 2 "out of place...since it applies to the abode of El after he has been ousted from heaven." Hence, he concluded that the occurrence of the phrase in v. 2 refers to Tyre, whereas its occurrence in v. 8 refers to the abode of El. This interpretation downplays virtually everything in the oracle but the phrase *ʾănî ʾēl* and the banishment in v. 8. It misses what I believe is the more salient aspect of the underlying myth, the juxtapositioning of the claim, *ʾănî ʾēl* 'I am ʾēl', with the counterassertion, *wəʾādām ʾattâ* 'but you are *ʾādām*'.

27, Tyre metaphorically *lives* (v. 4) and *dies* (vv. 26–27; cf. 26:5) 'in the heart of the seas'. The same, then, is most likely true in chapter 28, according to which 'in the heart of the seas' expresses both the dwelling of the protagonist (v. 2) and the place of his death (v. 8). While *bəlēb yammîm* seems to be an explanatory gloss for *môšab ʾĕlōhîm*, it does not belong *exclusively* to the myth complex upon which 28:2–10 is constructed.[378] As I pointed out in chapter 2, the connection of water and the divine habitation is evident throughout the ancient Near East.[379] Its occurrence in chapter 28 is most likely an *intentional* and subtle mixing of metaphors which is effective if one keeps both chapters in mind.

One suggestion might be made concerning a possible mythic background to the phrase 'in the heart of the seas'. In Mesopotamian traditions, the dwelling of Enki/Ea was the Apsu, the subterranean sweet waters under the earth. In chapter 2 we discussed the references to other gods in the ancient Near East being located at similar water sources: El-Kunirsa at the headwaters of the Euphrates river, and Ugaritic El at the 'sources of the (two) rivers in the midst of the streams of the two deep places'. Most significant for our purposes is the fact that El was identified with Enki/Ea in antiquity. The tradition of this identification is preserved in Berossus's account of the Flood, which substitutes Kronos, the Greek equivalent of El, in the place of Enki/Ea.[380] This is important because Enki/Ea was the god of wisdom,

---

[378]Further evidence of this may be seen in Zech 9:4, where the prophet says concerning Tyre, 'Yahweh will strip her of her possessions and hurl her wealth into the sea, and she shall be devoured by fire'. Tyre's demise in its ocean venue was apparently not an uncommon image.

[379]The connection of the phrase *bəlēb yammîm* with the dwelling of El has also been made by Clifford (*The Cosmic Mountain*, 170), who states that the phrase is "the Hebrew equivalent of Ugaritic *qrb ʾapq thmtm*," the dwelling of El. On the basis of Mesopotamian references to the removal of flood heroes to the divine habitation located at the source of the rivers, it is not necessary to understand *bəlēb yammîm* as referring *exclusively* to the abode of El. The fact that the text here refers to it explicitly as *môšab ʾĕlōhîm* (perhaps a conscious avoidance of the designation *ʾēl*) makes this conclusion even more reasonable.

[380]Berossus begins the account: 'Cronus appeared to Xisouthros [= Sumerian Ziusudra] in a dream and revealed that on the fifteenth day of the month Daisios mankind would be destroyed by a flood' (translation of S. Burstein, *Berossus the Chaldean* [Sources from the Ancient Near East 1,5; Malibu, Calif.: Undena, 1978], 20). El is identified with Kronos by Philo of Byblos (for text see H. Attridge and R. Oden,

as was also true for El in the Ugaritic literature. As we have noted and will continue to discuss, this oracle of Ezekiel and that which follows place considerable emphasis on the wisdom of the protagonist. It is not unreasonable to detect in the reference to a dwelling of the gods in the heart of the seas a reflection of the ancient Near Eastern wisdom traditions surrounding Enki/Ea, Oannes/Adapa (and the seven sages), and El. In the chapter 3 we discussed the general notion of these water sources being the place where gods dwell, and the place to where, as in the case of Utnapishtim, humans who become like gods are removed to eternal life. In the Israelite materials, we observed a similar understanding of subterranean waters in the garden account, and how these waters are connected with the temple (the divine habitation) and interpreted as the "fountain of life." We can add to this references from the Israelite wisdom tradition, based, no doubt, on these notions. Prov 16:22 says, 'insight (*śēkel*) is a fountain of life (*məqôr ḥayyîm*) to him who has it' (Prov 13:14, cf. 10:11,13). Likewise, according to Prov 18:4, 'the words of a man's mouth are deep waters, a gushing stream (*naḥal nōbēaᶜ*) is the fountain of wisdom (*məqôr ḥokmâ*)'.[381] The matchless wisdom of the protagonist, coupled with the association of wisdom with the subterranean cosmic waters in the ancient Near East provides a plausible motive for Ezekiel's linking Tyre's epithet 'in the heart of the seas' with 'the dwelling of the gods'.

In our examination of Genesis 2–3, we discussed the possible connection between the Israelite primal human and Mesopotamian Adapa/Oannes, who was known to come from the sea. There is also a tradition in the Erra Epic according to which Marduk banishes the sages (of whom Adapa is traditionally the first) saying, 'I dispatched those (renowned) *ummānū* (sages) down into the Apsu: I did not or-

---

Philo of Byblos: The Phoenician History, 20). For a discussion of interpretive issues surrounding such identifications, see R. Mondi, "Greek Mythic Thought in the Light of the Near East," *Approaches to Greek Myth* (ed. L. Edmunds; Baltimore: Johns Hopkins University Press, 1990), 141–97.

[381]Similar to these is Prov 3:18, where wisdom is 'a tree of life (*ᶜēṣ ḥayyîm*) to those who lay hold of her'. For discussion of the connection of wisdom literature to mythology, and references, see W. McKane, *Proverbs: A New Approach* (London: SCM Press, 1970), 296.

dain their coming up again'.[382] These are admittedly remote parallels, but given Ezekiel's connection to the Babylonian cultural milieu within the context of the Exile, this degree of contact is not impossible.

Much confusion is eliminated when the phrase 'in the heart of the seas' is considered a less central aspect of what Ezekiel wishes to say. This decision makes manifest what I believe is a clearer picture of the myth complex to which Ezekiel alludes:

> Because you have said "I am God,
> I dwell in the dwelling of the gods" —
> but you are human and not God,
> though you have made your mind like the mind of gods....

We are left here with what is unmistakably reminiscent of the myth of the primal human as preserved in the opening chapters of Genesis; this line also has a fairly clear analogue in the opening statement of the second oracle. Both oracles then offer a statement about the character of the protagonist and about his location.

Wisdom and conflict

The phrase *wattittēn libbəkā kəlēb ʾĕlōhîm* 'You have made your mind like the mind of gods' (v. 2, cf. v. 5) is ambiguous. It is universally agreed that *lēb* here pertains to the seat of the intellect, and that the phrase, therefore, pertains to divine knowledge or *wisdom*. It is not immediately clear, however, whether the protagonist *was* as wise as gods or merely *thought* himself to be. RSV translates 'though you consider yourself as wise as a god'. The use of the verb *ntn*, however, suggests more literally 'You have *made* your mind *like* the divine mind', suggesting some action and attendant result. In favor of the understanding "thinking oneself to be wise" is the fact that the phrase seems to be related (or parallel) to *gābah libbəkā* in v. 2: each is introduced by the conjunction *yaʿan* (vv. 2, 5). Still, vv. 3–5 suggest in no

---

[382]Tablet I, line 148; translation of L. Cagni, *The Poem of Erra* (Sources and Monographs on the Ancient Near East 1,3; Malibu, Calif.: Undena, 1977), 32. Note that the second tablet deals with Ea's attempt to influence Marduk to return the sages from the Apsu (lines 16–20). In addition to Cagni, see F. N. H. Al-Rawi and J. Black, "The Second Tablet of Išum and Erra," *Iraq* 51 (1989): 111–22. Apsu is the cosmic deep, and the dwelling of the god of wisdom, Enki.

uncertain terms that the protagonist possesses supernal wisdom ('no secret is hidden from you'), and these verses seem quite clearly to have been set in to explain *wattittēn libbəkā kəlēb ʾĕlōhîm* which immediately precedes it. It is not unreasonable to see in this passage a reflection of the statement of God in Gen 3:22: *hāʾādām hāyâ kəʾaḥad mimmennû lādaᶜat ṭôb wārāᶜ* 'The man has become *like one of us*, knowing good and evil'.

The punishment for the primal human's misdeed is expressed in clever fashion by the prophet. He is told in v. 7 that strangers will 'defile his splendor' (*wəḥilləlû yipᶜātekā*) and that:

לַשַּׁחַת יוֹרִדוּךָ וָמַתָּה מְמוֹתֵי חָלָל בְּלֵב יַמִּים

They will cast you down to the pit, and you will die the death of those who are profaned in the heart of the seas.

We can be assured that 'defiled' is a better translation than 'slain' or 'pierced' on two accounts. First, in the parallel image in v. 10 we read, 'You will die the death of the *uncircumcised*' (*môtê ᶜărēlîm tāmût*). Second, it is precisely the same image used by Ezekiel in the second oracle: 'I cast you as a profane thing from the mountain of God' (*ʾăḥallelkā*; 28:16, literally, 'I *profaned* you'). What is more, this 'death of the uncircumcised' in 28:10 — another unambiguous reference to the unclean and profane — will be carried out by 'those who will defile you' (*məḥaləleykā*).[383] Here, *bəlēb yammîm* functions not only as a description of historical Tyre or the divine abode; it is also a synonym for Sheol. The very text we are presently examining is evidence of the use of the phrase for the divine abode. That the phrase can also refer to Sheol is clearly seen in Jonah 2:2–4:

[2]וַיִּתְפַּלֵּל יוֹנָה אֶל־יְהוָה אֱלֹהָיו מִמְּעֵי הַדָּגָה׃ [3]וַיֹּאמֶר קָרָאתִי מִצָּרָה לִי
אֶל־יְהוָה וַיַּעֲנֵנִי מִבֶּטֶן שְׁאוֹל שִׁוַּעְתִּי שָׁמַעְתָּ קוֹלִי׃ [4]וַתַּשְׁלִיכֵנִי מְצוּלָה
בִּלְבַב יַמִּים וְנָהָר יְסֹבְבֵנִי כָּל־מִשְׁבָּרֶיךָ וְגַלֶּיךָ עָלַי עָבָרוּ׃

[2]Jonah prayed to Yahweh his god from inside the fish, [3] and he said, "I cried out in my distress to Yahweh, and he answered me. Out of the

---

[383]In vv. 7–10, death and defilement (or profanation) are paired. Verse 7: 'they will draw their swords'/ 'they will defile'; v. 8: 'they shall thrust you into the pit' / 'death of the defiled'; v. 9 'those who slay you'/ 'those who defile'; v. 10 'you will die the death of the uncircumcised'.

> belly of Sheol I cried; you heard my voice. [4]You had cast me (*šlk*) into the deep, into *the heart of the seas* (*bilbab yammîm*). The river surrounded me, all your waves (*mišbāreykā*) and your breakers passed over me."

Ps 88:4–8 also reflects a strong tradition equating Sheol with the sea. Verses 7 and 8 read:

> 7 שַׁתַּנִי בְּבוֹר תַּחְתִּיּוֹת בְּמַחֲשַׁכִּים בִּמְצֹלוֹת׃
> 8 עָלַי סָמְכָה חֲמָתֶךָ וְכָל־מִשְׁבָּרֶיךָ עִנִּיתָ סֶּלָה׃

> 7 You have put me in the depths of the Pit, in regions dark and deep.
> 8 Your anger rests upon me, and you overwhelm me with all your waves.

Thus, the primal human is cast out of the divine habitation into Sheol, an image both of expulsion and of death, the same fate, essentially, as that of the first couple in Genesis.

6. Personified Wisdom and the Culture-Bearer Tradition (Proverbs 8:22–31)

PROVERBS 8:22–31 PRESENTS A TEXT in which the primal human tradition is greatly sublimated, yet, in the end, clearly present. A cursory reading suggests nothing immediately reminiscent of the primal human sphere of ideas. In this passage the poet-sage writes of a completely new figure, wholly other, a feminine being simply called "Wisdom." Moreover, the fact that the figure is portrayed in the feminine suggests even further to the casual observer that there is no connection to the primal human concept. But placed within the context of what we have already examined, the connections are striking. What we find ultimately is a figure who shares in a number of traditions — not the least of which is that of the primal human conception.

The Text: Proverbs 8:22–31

22 יְהוָה קָנָנִי רֵאשִׁית דַּרְכּוֹ קֶדֶם מִפְעָלָיו מֵאָז׃
23 מֵעוֹלָם נִסַּכְתִּי מֵרֹאשׁ מִקַּדְמֵי־אָרֶץ׃
24 בְּאֵין־תְּהֹמוֹת חוֹלָלְתִּי בְּאֵין מַעְיָנוֹת נִכְבַּדֵּי־מָיִם׃
25 בְּטֶרֶם הָרִים הָטְבָּעוּ לִפְנֵי גְבָעוֹת חוֹלָלְתִּי׃
26 עַד־לֹא עָשָׂה אֶרֶץ וְחוּצוֹת וְרֹאשׁ עַפְרוֹת תֵּבֵל׃
27 בַּהֲכִינוֹ שָׁמַיִם שָׁם אָנִי בְּחוּקוֹ חוּג עַל־פְּנֵי תְהוֹם׃
28 בְּאַמְּצוֹ שְׁחָקִים מִמָּעַל בַּעֲזוֹז עִינוֹת תְּהוֹם׃
29 בְּשׂוּמוֹ לַיָּם ׀ חֻקּוֹ וּמַיִם לֹא יַעַבְרוּ־פִיו בְּחוּקוֹ מוֹסְדֵי אָרֶץ׃
30 וָאֶהְיֶה אֶצְלוֹ אָמוֹן וָאֶהְיֶה שַׁעֲשֻׁעִים יוֹם ׀ יוֹם מְשַׂחֶקֶת לְפָנָיו בְּכָל־עֵת׃
31 מְשַׂחֶקֶת בְּתֵבֵל אַרְצוֹ וְשַׁעֲשֻׁעַי אֶת־בְּנֵי אָדָם׃

22 Yahweh created me at the beginning of his way,
Before his deeds of old.
23 Long ago *I was installed*,
From the beginning, from before the earth.
24 When there were no depths, *I was brought forth*,
when there were no streams heavy with water.

[25] Before the mountains were set in,
before the hills, I was brought forth.
[26] When he had not yet made the earth and the outer parts,
the beginning of the dust of the world.
[27] When he established the heavens I was there.
when he inscribed a circle on the face of the deep,
[28] When he made firm the clouds above,
when he strengthened the springs of the deep,
[29] When he set for the sea his decree
that the waters would not transgress his command,
When he inscribed the foundations of the earth —
[30] I was beside him, an *ʾāmôn;*
I was delighting daily,
rejoicing before him continually,
[31] Rejoicing in his inhabited world,
My delight was with humanity.

In this chapter we shall draw attention to some of the similarities between the discussion of the primal human in Job and that concerning Wisdom in Prov 8:22–31. The verbatim correspondence between these passages has been before the attention of commentators for many years, but the relationship between the two texts is not agreed upon. Similarities notwithstanding, the texts deal with two different figures: so-called "Dame Wisdom" in Proverbs, and the primal human in Job. Nonetheless, analysis of the texts reveals a connection and demonstrates the concept of the culture bearer as a central link between the two.

Earlier commentators studying the correspondence between the two passages posited a basis in a myth about a preexistent "primordial man," for which little direct evidence existed. Schenke, for example, wrote, "es ist vielleicht nicht zu kühn anzunehmen, daß Prv. 8:24 f. einer Schilderung des präexistenten Urmenschen entnommen ist, aus der man sowohl für Hi. 15 als für Prv 8 geschöpft hat."[384] The common pool of images that Schenke suggests surrounding this primordial being in the Israelite period remains difficult to establish without appealing to some external paradigm. We have already discussed the difficulty in positing the existence of such a preexistent figure on the basis of evidence in the Hebrew Bible alone. Still, on examination of

[384] W. Schenke, *Die Chokma (Sophia) in der jüdischen Hypostasenspekulation.* (Videnskapsselskapts Skrifter II Hist.-Filos. Klasse 1912. No 6; Kristiana: Utgit for H. A. Benneches Fond, in Kommission bei J. Dybwad, 1913), 17.

Proverbs 8 and Job 15, it seems best to conclude that Prov 8:22–31 also borrows the appropriate language of the primal human and incorporates it into the larger framework of the discussion of the figure Wisdom.[385] It is certainly not the case that Eliphaz is referring to the figure *ḥōkmâ*, for he expressly draws the comparison with *first human* (*rîˀsôn ˀādām*), and ascribes to the primal human the *attribute* of wisdom (or posits wisdom as a *possession*). The association is very close, however, for the point he makes requires wisdom to be, in fact, the primary defining characteristic (or possession) of primal human.

Wisdom and the Primal Human

In placing the texts side by side (that is, Prov 8:22–31 and the texts discussed above concerning the primal human in the book of Job), one discovers points of contact between the texts that go beyond what scholars generally recognize. This discovery may be summarized as follows: Prov 8:22–31 includes three elements which emerge from the picture of the primordial human presented in the book of Job: a) a verbatim reference to the subject's origins; b) a series of temporal clauses indicating the subject's preexistence; c) a reference to jubilation at the creation.

The subject's origins

The verbatim correspondence with regard to the subject's origins constitutes the most obvious and striking link. We recall in Job 15:7, Eliphaz asks Job: *hărîˀšôn ˀādām tiwwālēd wəlipnê gəbāᶜôt ḥôlāltā* 'Were you born the first human? Were you brought forth before the hills?'. Likewise, Wisdom here says, *lipnê gəbāᶜôt ḥôlāltî* 'Before the hills, I was brought forth' (Prov 8:24). The imagery employed is that of birth.

The meaning of this reference cannot be adequately understood without first examining that which precedes it in v. 22: 'Yahweh

[385]Cf. H. May ("The King in the Garden of Eden," 171), who understood that wisdom in Prov 8:22–31 "is personified in terms of primordial First Man." Cf. H. Ringgren, ***Word and Wisdom: Studies in the Hypostatization of Divine Qualities and Functions in th Ancient Near East*** (Lund: Hākan Ohlssons Boktryckeri, 1947), 90.

created me at the beginning of his way'. What is intended in the verb *qānānî* is directly relevant, since it is not clear whether the verb here means 'created', or 'begotten', or 'acquired'. De Boer has argued that Wisdom already existed when God began his creative activity, and therefore Yahweh simply appropriated it into his work.[386] Ringgren (following Nyberg) recognized the closeness of the two meanings 'to possess' and 'to create'. As a result he found no substantial difference between them, since one may acquire a thing by a number of means, one of which (in the case of a child) is begetting.[387] The fact that the verb *ḥôlaltî* appears makes it more likely that birth imagery is involved. Nevertheless, as we discussed in chapter 4, creation and birth imagery is often mixed. The clear idea in this passage as it stands is that Wisdom existed before the creation; the mode by which it came into existence is less clear.

Another clue to the manner of Wisdom's emergence is found in the term *nissaktî*, 'I was appointed'. Feuillet considered this a reference to the royal investiture of the being and cited Ps 2:6, 'I have set up (*nāsaktî*) my king on Zion, my holy hill'.[388] I agree, and find that this linguistic link to royal investiture strengthens the case for interpreting the poem as imagery borrowed from primal human.

Temporal clauses

There is a clearly demonstrable connection between the texts pertaining to the primordium in Job 38 and Prov 8:22–31 in the series of temporal clauses. These clauses in vv. 22–31 attracted the attention of B. Gemser, who found the poem reminiscent of the literary style known in the ancient Near East, the "creation hymn." He, and H. Ringgren, who concurred, cited as parallels sections from the Egyptian Book of Apophis, the Babylonian Enuma Elish, the bilingual text known as "The Creation of the World by Marduk," and the two

---

[386]P. A. H. de Boer, "The Counsellor," in *Wisdom in Israel and the Ancient Near East* (ed. M. Noth and D. W. Thomas; VTSup 3; Leiden: E. J. Brill, 1969), 42–71.

[387]H. Ringgren, *Word and Wisdom*, 101.

[388]A. Feuillet, "Le fils de l'homme de Daniel et la tradition biblique," *RB* 60 (1953) 323–24; cf. May, "The King in the Garden of Eden," 172.

accounts of creation in Genesis — 1:1–3 and 2:4b–7.[389] The passage from Enuma Elish may be cited here as an example:

> When on high the heaven had not been named,
> Firm ground below had not been called by name,
> Naught but primordial Apsu, their begetter,
> (And) Mummu-Tiamat, she who bore them all,
> Their waters commingling as a single body;
> No reed hut had been matted, no marsh land had appeared,
> When no gods whatever had been brought into being,
> Uncalled by name, their destinies undetermined —
> Then it was that the gods were formed within them.[390]

R. N. Whybray, however, has argued vigorously against interpreting any meaningful association among the various texts, finding that they

> are of various kinds and hardly comparable one with another except for the simple facts that they all refer to the creation of the world and that in doing this they (in most cases at least) employ similar grammatical constructions: negative temporal clauses. Neither in their general purpose nor in the function of these temporal clauses can any strong similarities be found.[391]

Hence, Whybray finds no relevance in the similarities of these texts to cast light on the issue of wisdom. The grammatical parallels among these texts are indeed striking, and perhaps Whybray has understated their significance.[392] But he does make a reasonable point in calling for greater scrutiny in assessing the significance of the similarities. I believe the primal human allusions in the book of Job are the most appropriate context within which to set these verses in Proverbs 8.

In chapter 4, I pointed out the importance of the speeches of the

---

[389]B. Gemser, ***Sprüche Salamos*** (HAT 16; Tübingen: J. C. B. Mohr, 1969), see 44–49. H. Ringgren, ***Word and Wisdom***, 102.

[390]***EnEl*** I 1–9. Speiser's translation, ***ANET***, 60–61.

[391]R. N. Whybray, "Proverbs 8:22–31 and Its Supposed Prototypes," *VT* 15 (1965): 513.

[392]Whether the similarities are best understood as "coincidences," as Whybray seems to contend, is debatable. Cf. the statement (ibid.), "we must constantly pay due attention to the possibility of coincidence. Any account of the creation of the world is likely, in view of the nature of the subject, to have some points of resemblance with others — e.g., one would expect references to heaven, earth, plants, living creatures, man, etc."

theophany to the understanding of Eliphaz's reference to the primal human in 15:7. Eliphaz had argued Job's lack of wisdom concerning his situation in spite of Job's statements to the contrary. To make his point, Eliphaz set up the primal human as a foil to Job. The only way Job could claim to have higher wisdom than Eliphaz would be if he were the primal human himself; and in Eliphaz's estimation, Job was no primal human. The book centers around the subject of divine insight, the knowledge of things beyond the scope of the normal human experience. I pointed out that Eliphaz's question, 'What do you know?' (15:9), was directly related to his question regarding whether Job was the primal human, born in the primordium. As we saw, this same combination of ideas, *knowing* and being *born in the primordium* reappeared in the speech of the theophany in 38:21: 'You know, for you were born then, and the number of your days is great'.

In this section God asks Job a series of questions, which recalls the encyclopedic knowledge of those claiming to be wise. In 38:4–11 the following series of questions appears:

4 אֵיפֹה הָיִיתָ בְּיָסְדִי־אָרֶץ הַגֵּד אִם־יָדַעְתָּ בִינָה׃
5 מִי־שָׂם מְמַדֶּיהָ כִּי תֵדָע אוֹ מִי־נָטָה עָלֶיהָ קָּו׃
6 עַל־מָה אֲדָנֶיהָ הָטְבָּעוּ אוֹ מִי־יָרָה אֶבֶן פִּנָּתָהּ׃
7 בְּרָן־יַחַד כּוֹכְבֵי בֹקֶר וַיָּרִיעוּ כָּל־בְּנֵי אֱלֹהִים׃
8 וַיָּסֶךְ בִּדְלָתַיִם יָם בְּגִיחוֹ מֵרֶחֶם יֵצֵא׃
9 בְּשׂוּמִי עָנָן לְבֻשׁוֹ וַעֲרָפֶל חֲתֻלָּתוֹ׃
10 וָאֶשְׁבֹּר עָלָיו חֻקִּי וָאָשִׂים בְּרִיחַ וּדְלָתָיִם׃
11 וָאֹמַר עַד־פֹּה תָבוֹא וְלֹא תֹסִיף וּפֹא־יָשִׁית בִּגְאוֹן גַּלֶּיךָ׃

4 Where were you when I laid (*bəyosdî*) the foundation of the earth?
Tell me if you have understanding.
5 Who determined (*śām*) its measurements?
Surely you know!
Or who stretched the line upon it?
6 On what were its bases sunk,
Or who laid its cornerstone,
7 When the morning stars sang (*bəron-*) together
And all the sons of God shouted for joy?
8 Who shut in the sea with doors
When it gushed forth (*bəgîḥô*) from the womb,
9 When I made (*bəśûmî*) a cloud its garment,
Darkness, its swaddling band,
10 And I set upon it my bounds (*ḥuqqî*)

And set (*wāʾāśîm*) bars and doors
11 And said, "You may come to this point but no farther,
And at this point your proud waves shall stop"?

We have already seen in chapter 4 how this text and 38:21 suggest the presence of the primal human in the primordial situation. There is yet another observation to be made regarding this text and Prov 8:22–31. The Job 38 passage is governed by the initial question, 'Where were you?'. The knowledge the passage speaks of presupposes presence in that situation. Prov 8:22–31 seems to be a rather emphatic answer to this question, 'I was there'. Although the entire passage might be seen as addressing this concern, vv. 27–30 speak most strikingly:

27 When he established (*bəhăkînô*) the heavens I was there (*šām ʾănî*)
when he inscribed (*bəḥûqô*) a circle on the face of the deep.
28 When he made firm (*bəʾamməṣô*) the clouds above,
when he strengthened (*baʿăzôz*) the springs of the deep.
29 When he set (*bəśûmô*) for the sea his decree (*ḥuqqô*).
that the waters would not transgress his command.
When he inscribed (*bəḥûqô*) the foundations of (*môsədê*) the earth.
30 I was beside him, an *ʾāmôn*.

This series of temporal clauses (the preposition *bə-* prefixed to the infinitive construct) is positive and addresses Yahweh's concern in the question, 'Where were you *when* I...' (*bə-* plus infinitive construct). Particularly forceful in this regard are the assertions of vv. 27 and 30, respectively: '*I was there*' (*šām ʾănî*) and '*I was beside him*' (*wāʾehyeh ʾeṣlô*). The correspondence between Prov 8:29 and Job 38:8–11 where the limit is placed on the sea is evident as well.

"Jubilation" at the creation

The reference to the sons of God rejoicing at the creation in Job 38 has been discussed above; and the connection to the idea of the primal human in Job has also been addressed. This theme apparently reappears in Prov 8:30–31, the meaning of which has proven to be rather elusive. Verses 30aβ–31 read:

I was delighting (?) daily,
rejoicing before him continually,
rejoicing in his inhabited world,
my delight was with humanity.

Commentators have had great difficulty making sense of the difficult language of this passage, particularly with respect to the context within which it is set. Given other correspondences with Job, it is reasonable to seek an interpretation based on the concept of rejoicing at the creation, as in Job 38:4–7.[393]

Wisdom and the Culture-Bearer Tradition

The similarity of Wisdom and the primal human here suggests another aspect of the primal human: his role as culture-bearer, based on the use of the term *ʾāmôn*.

A frequently discussed issue is whether the figure described in Prov 8:22–31 played an active role in the creation. Central to the discussion is the disputed meaning of the term *ʾāmôn*. Some commentators accept the meaning 'child' or 'ward', based on the meaning of the root *ʾmn* 'to support, nourish'. Kayatz understood the figure to be comparable to the Egyptian figure Maat, who is in certain texts represented as a child who plays before and delights the gods.[394] Similarly Gemser, disagreeing with the interpretation that Wisdom played a role in the creation, writes, "it is only said of Wisdom that she was present at the creation, not as a helper, but like a child, who played in the workshop of her father."[395]

In a well-known article, P. A. H. de Boer argued that the word should be read *ʾimmōn*. He believed the term to be related to *ʾēm* 'mother', which he had attempted to establish as an official title of a

---

[393]According to A. Ross, Prov 8:30–31 "tells how wisdom rejoiced at God's creation." Ross, however, mentions no connection with Job 38:4–7. Instead he relates it to Genesis 1, and translates, 'I was filled with delight day after day, rejoicing in his presence, rejoicing in his whole world and delighting in mankind'. The phrase "delight day after day," he writes, "recalls that 'God said it was good' Genesis 1, *passim*." A. Ross, "Proverbs," in *The Expositor's Bible Commentary* (ed. F. Gaebelein; Grand Rapids: Zondervan, 1991), 5.946. There is insufficient evidence to demonstrate a direct link between this passage and the creation account of Genesis 1; nevertheless, the suggestion is intriguing. Whether Ross makes the connection between *yôm yôm* and *yôm ʾeḥād yôm šēnî* is not clear. This connection was brought to my attention by J. Hackett.

[394]C. Kayatz, *Studien zu Proverbien 1–9* (WMANT 22; Neukirchen-Vluyn: Neukirchener Verlag, 1966), 93–94.

[395]B. Gemser, *Sprüche Salamos*, 47.

counselor. De Boer offered as possible meanings 'mother-official', which he regarded a "descriptive term for the function of counsellor"; or the diminutive 'little mother', which he believed fit equally well with the context.[396] In her role as advisor, Wisdom served in the capacity of the queen mother.[397]

The term *ʾāmôn* appears only here and in Jer 52:16. The context of Jer 52:16 suggests a meaning of 'artisan' or 'craftsman'. The understanding of the term as 'craftsman' is reflected in Wisd 7:22 [Gk 21], where Wisdom is called ἡ πάντων τεχνῖτις 'the fashioner of all things'.[398] Second Kings 25:11, the parallel passage to Jer 52:16, reads *hāmôn* ('crowd'), but it is not clear whether this is the original reading or reflects another form of the same word.[399] W. F. Albright considered it a Canaanite reminiscence and translated it "craftsman, master artificer, wizard, etc."[400]

Smith, following Zimmern, traced the word back to Akkadian *ummānu*. The Mesopotamian *ummānu* was, as we have seen, a type of scribe or sage. Often it refers to a particular kind of scribe, more expert than the normal *ṭupšarru* scribe. The term refers to an official position in both the Assyrian and the Babylonian administrations.[401] Gaster accepted the connection to Akkadian *ummānu* and suggested that it implies the role of a court expert.[402] Oppenheim saw a corre-

---

[396]P. A. H. de Boer, "The Counsellor," 42–71.

[397]See McKane for a discussion of the problems with de Boer's position, particularly the assertion of the diminutive 'little mother' (W. McKane, *Proverbs: A New Approach* (London: SCM Press, 1970), 356–57.

[398]The understanding 'craftsman' is also supported by LXX, Vulgate, Syriac, and Targum. A related idea appears in *Gen. Rab.* 1:1, where it is applied to Torah.

[399]According to S. Smith, the form *hāmôn* in 2 Kgs 25:11 is a case of variation normally expected with loanwords. He likewise considered *ʾāmān* in Cant 7:2 a variant form of the loanword, and thus read, 'the swayings of your hips are like those of ornaments made by a master-craftsman's hands' ("An Inscription from the Temple of Sin at Huraidha in the Hadhramawt," *BSOAS* 11 [1945]: 456).

[400]W. F. Albright, "Some Canaanite-Phoenician Sources of Hebrew Wisdom," VTSup 3 (1969): 8. Albright based his translation 'wizard' on an appearance of the word in the Taanach tablets. See also *BASOR* 94 (1944): 18 n. 28.

[401]Smith, "An Inscription," 457.

[402]T. Gaster *VT* 4 (1954): 77. Cf. E. Reiner's assertion that the *apkallu* represents a vizier ("The Etiological Myth of the Seven Sages," 1–11).

spondence with Mummu, the vizier of Apsu in Enuma Elish, where Apsu is quoted as saying, 'O Mummu, vizier, who brings me into a good mood'.[403]

This interpretation of *ʾāmôn* is significant for our purposes, for it reintroduces the figure Adapa and the Mesopotamian culture-bearer tradition into the discussion. We have discussed the culture-bearer tradition in chapter 2, and the significance of the terms *ummānu* and *apkallu* with respect to it. The interest the figure expresses in humanity in Prov 8:31 reflects that of the culture-bearer, and the relationship between God and Wisdom may be likened to that of the king and his *ummānu* or *apkallu*. The subject is posited as an intermediary figure between God and humanity, and the language of the passage is highly suggestive of the primal human. Hence, we may conclude that in the background of what in its present context is a poem about "Dame Wisdom," the primal human is presented in the capacity of culture-bearer, a figure who brings divine knowledge to humanity. We should not find it surprising that such imagery is appropriated by a figure whose chief concern is that humanity should have wisdom.

---

[403]*EnEl* I 31; *ANET* 61.

7. Cain and Noah: Other Manifestations of Primal Human Imagery

THE EXTENSION OF PRIMAL HUMAN IMAGERY in the Hebrew Bible is found in other subtle ways that reveal an interest in the primal human that again transcends any simple "historical" conception of Adam. Let us now briefly return to the book of Genesis to examine two such narratives. These examples are introduced within the context of the primeval history and consciously reiterate previously stated literary elements associated with Adam.[404] The two figures involved are Cain and Noah.[405] To be sure, these are distinct figures, individuals in their own right, but the fact that they, like Adam, are antediluvian figures prepares us for a degree of similarity. The connections are strong enough to warrant locating these figures on the periphery of the primal human traditions.

Cain and the Primal Human Tradition

There are numerous parallels between the story of Adam in Genesis 3 and that of Cain in Genesis 4. I wish only to point out those of interest to our study.[406] It is logical that Cain might be presented with primal human imagery, for he is the first human born of a woman. The circumstances surrounding his birth are somewhat mysterious. Gen 4:1 relates the event in the following words:

---

[404]An analogous literary situation may be found within the Šemiḥaza narrative in 1 Enoch 6–11. P. D. Hanson ("Rebellion in Heaven, Azazel, and Euhemeristic Heroes in 1 Enoch 6–11," *JBL* 96 [1977]: 195–233) refers to the Azazel episode as an "expository narrative" and an "interpretive elaboration growing organically out of the Šemiḥaza narrative." The narrative picks up on original elements in the text and combines them with other mythological elements.

[405]For a treatment of possible parallels with the Mesopotamian Atrahasis Epic, see I. Kikawada and A. Quinn, *Before Abraham Was: The Unity of Genesis 1–11* (Nashville: Abingdon, 1985), 36–53.

[406]For a discussion of the vast literature on the similarities between the two stories, see Westermann, *Genesis*, 282–320 and references.

וְהָאָדָם יָדַע אֶת־חַוָּה אִשְׁתּוֹ וַתַּהַר וַתֵּלֶד אֶת־קַיִן וַתֹּאמֶר קָנִיתִי אִישׁ אֶת־יְהוָה׃

The man knew Eve his wife and she conceived and bore Cain. And she said "I have gotten a man with Yahweh."[407]

There is ambiguity in the verb *qānâ*, which can also mean 'to create'.[408] Thus, despite the apparent natural means of birth, the woman seems also to say, *'I have created a man with Yahweh'*.

Cain is *ᶜōbēd ᵓădāmâ* 'a worker of the soil' (Gen 4:1), recalling that the original primal human was placed in the garden *ləᶜobdāh ûləšomrāh* 'to work it and keep it'. It is interesting that he is juxtaposed with his brother Abel (*hebel*) who is a *rōᶜēh ṣōᵓn* 'a shepherd of sheep' (Gen 4:2). In chapter 2 we located the term shepherd in the same complex of ideas according to which the Mesopotamian king could bear the epithet *nukaribbu* 'gardener', and we saw that gardeners figured prominently in mythological texts. Cain is also a "culture bearer" in Hebrew tradition; we read in 4:17 that he went on to become the (first) builder of a city.[409]

Noah and the Primal Human Tradition

Primal human imagery reemerges in the figure of Noah. Noah is a primal human in the sense that all humans descend from him, following the destruction of the flood.[410] Noah is called a righteous man in Gen 6:9. We also read *ᵓet–hāᵓĕlōhîm hithallek-nōaḥ* 'Noah walked with God'. This exact statement was made twice concerning Enoch in Gen 5:22,24. Verse 24 reads *wayyithallēk ḥănôk ᵓet–hāᵓĕlôhim* 'Enoch

[407] I. Kikawada has connected this reference to the "double creation" of humanity in the Sumerian myth "Enki and Ninmah" ("Two Notes on Eve," *JBL* 91 [1972]: 33–37; "The Double Creation of Mankind in Enki and Ninmah, Atrahasis I 1–351, and Genesis 1–2," *Iraq* 45 [1983]: 43–45. See also I. Kikawada and A. Quinn, *Before Abraham Was: The Unity of Genesis 1–11*, 39–40).

[408] The root *qnh* is used with the meaning 'to create' several times with reference to God (cf. especially Gen 14:19; Deut 32:6; Ps 139:13).

[409] Cf. W. W. Hallo, "Antediluvian Cities," *JCS* 23 [1970]: 64). Perhaps kingship is implied in this act as well.

[410] On the first human as flood hero, see A. Kilmer, "Speculations on Umul, the First Baby," in *Kramer Anniversary Volume: Studies in Honor of Samuel Noah Kramer* (AOAT 25; ed. B. L. Eichler et al. [Kevelaer: Butzon & Bercker,1976]): 265–70.

walked with God'. As is well known, he apparently escapes death, for the text continues, *wəʾênennû kî lāqaḥ ʾōtô ʾĕlōhîm* 'and he was not (or, disappeared) for God took him'. The phrase "walking with (*hlk* Hithpael) God" signifies a special relationship to deity. It recognizes the humanity of the individual while placing that person in a non-human context. In chapter 3 I argued that the statement, 'you were on the holy mountain of God; among stones of fire you walked', in Ezek 28:14 bore the same significance.

The genealogical account of the birth of Noah recorded in Gen 5:29 gives an etymology of his name which recalls both the story of *hāʾādām* in Genesis 2–3 but also the Mesopotamian understanding of the creation of humanity.[411] This section in question is considered by many to be a J addition to the Priestly genealogy. After the formulaic statement about Lamech's fathering a son, the text reads:

וַיִּקְרָא אֶת־שְׁמוֹ נֹחַ לֵאמֹר זֶה יְנַחֲמֵנוּ מִמַּעֲשֵׂנוּ וּמֵעִצְּבוֹן יָדֵינוּ מִן־
הָאֲדָמָה אֲשֶׁר אֵרְרָהּ יְהוָה׃

He called his name Noah, saying, "This one shall bring us relief from our work and from the toil of our hands, from the ground which Yahweh has cursed."

This apparent awareness of the Mesopotamian traditions about the creation of humanity supports the claim that the creation of *hāʾādām* and his appointment as gardener in Genesis 2 intentionally expressed *significant* information about the nature of the primal human and his relationship to God.

Recalling the language and imagery preserved in the Genesis 1 account of the creation of *hāʾādām*, Noah is blessed as was the primal human couple and given the same charge to 'be fruitful and multiply, and fill the earth' (9:1). The royal imagery is once again expressed in terms of humanity's domination of the earth. The statement that humanity is created in the image of God is reiterated in v. 6, where it is given as the reason for the prohibition against homicide. Finally, in 9:20 we read, *wayyāḥel nōaḥ ʾîš hāʾădāmâ* 'Noah was the first farmer'.[412]

---

[411]See above, chapter 2 in the discussion of humans as laborers of the gods.

[412]For discussion of the syntax of this verse see Westermann, *Genesis*, 481–82.

Conclusion: The Significance of Primal Human Traditions in Ancient Israel

WE BEGAN THIS INQUIRY WITH QUESTIONS concerning the significance of the primal human traditions in ancient Israel. We have examined just what those traditions are, that is, what constitutes them within the corpus of ancient Israelite literature.

We have seen that the primal human conception was a popular biblical notion that appeared in different modes and was put to a variety of uses. We have seen it presented directly and indirectly. We have witnessed it in the service of history[421] and basic aetiology (chapters 1 and 2) and have seen it used ahistorically to express analogies (chapters 3 and 4). Finally, we have observed sublimations of the conception in which it participates in the shaping of other ideas and images.

Through all of this, we have argued that the primal human is the significant forerunner of humanity who defines the relationship between humanity and deity. We conclude this study with a few additional observations about the significance of the primal human in Israelite society.

In exploring the meaning of myth, Mircea Eliade wrote:

> Any ritual whatever...unfolds not only in a consecrated space (i.e., one different in essence form profane space) but also in a "sacred time," "once upon a time" (*in illo tempore, ab origine*), that is, when the ritual was performed for the first time by a god, an ancestor, or a hero.[422]

Following the work of Gerardus van der Leeuw, he added:

> All religious acts are held to have been founded by gods, civilizing heroes, or mythical ancestors. It may be mentioned in passing about the primitives, not only do rituals have their mythical model but any human act whatever acquires effectiveness to the extent to which it exactly

---

[421]See above, p. 23.

[422]M. Eliade, *Cosmos and History: The Myth of the Eternal Return* (New York: Harper and Row, 1959), 22.

> *repeats* an act performed at the beginning of time by a god, a hero, or an ancestor.[423]

Eliade stressed the issue of efficacy in the actions based upon myths, finding that these various acts of repetition are "not merely a question of imitating an exemplary model"; rather, the "principal consideration is the result."[424] If Eliade was correct, then we should expect to find hints of the importance of the primal human beyond his being contained in a story. These ideas of Eliade's, I believe, provide considerable insight into the significance of primal human traditions in ancient Israel.

The texts we have examined use language and imagery that reveal the primal human to be a mediating or intermediary figure, standing between the human and divine realms. As such, the primal human conception emerges as a paradigm for intermediary figures in ancient Israelite society. It provided at least one stream of tradition in the construction and regulation of ideas concerning kingship and other offices of intermediation. There is a clear and consistent connection between biblical ideas on intermediation and the primal human texts we have examined in the past chapters.

The Primal Human Presented as Intermediary Figure

We have observed in our texts that the language used to describe the primal human is royal (Gen 1:26–28), priestly (Ezek 28:11–19), and prophetic (Job 15:7–8). What these roles share is that they are expressions of the role of intermediary. The primal human conception presents an image that is consonant with the notion of intermediation by virtue of the fact that he occupies a position between deity and humanity; moreover, he is the only one who can truly lay claim to that distinction. The primal human alone is not "born of woman." He is the only one whose natural state was face to face with God. He is the only one who lived in the "actual" (mythical) divine dwelling. Others can perform the function of intermediary, but in doing so they but mimetically follow the primal human. In this sense, the primal human is the

---

[423]Ibid.

[424]Ibid., p. 24.

significant ancestor who established the paradigm for contact with the divine.

The king mediates the divine will through just and proper rule.[425] He is a vice-regent and arm of the true king, the deity. The priest is a mediator first as an oracle-giver. He is also a mediator in his ability to stand before the presence of God and neutralize the encroaching forces of chaos through the cult — just as the primal human tended the garden. The prophet also stands before God and functions as messenger of divine will through proclamation. What all of these roles share in common is that they are defined in part by their ***positional significance***, an idea adumbrated in the person of the primal human: the very essence of their role as mediator comes from their "physical" proximity to God. All of these are mythically established in the location of the primal human in the divine abode.

Genesis: The Primal Human Presented as King

Of the two accounts of the creation of man in Genesis, Gen 1:26–28 unequivocally presents the primal human as king. Royal imagery is not as easily discerned in Genesis 2–3. It does appear, however, that the placing of the man in the garden to work and to keep it also has royal overtones. In chapter 2, these royal overtones were demonstrated in the royal interest in gardens in the ancient Near East — in Mesopotamian traditions as well as in Israelite literature. One of the royal aspects which appears to play a role in the language chosen by the J writer is that the Mesopotamian king could bear the epithet n u - k i r i$_6$ /*nukaribbu* 'gardener' or e n g a r /*ikkaru* 'farmer'. This is the most substantial link with kingship in the account of Genesis 2–3.

The Mesopotamian king could also bear the epithet s i p a or *rēʾû* 'shepherd'. Although we do not have evidence of the Israelite king bearing the epithet "gardener" or "farmer," there is abundant evidence of Israelite kingship being tied to the conception of the shepherd, particularly in the Davidic ideal of being taken from a life as a shepherd

[425]See T. H. Gaster, "Myth and Story," in ***Sacred Narrative: Readings in the Theory of Myth*** (ed. A. Dundes; Berkeley and Los Angeles: University of California Press, 1984), 120 (originally published 1954).

and raised to kingship. Even more persuasive, perhaps, is the fact that the term 'shepherd' is applied to Yahweh, who was frequently thought of in royal terms.[426] The fact that the god, the heavenly king, bears the epithet suggests that the earthly vice-regent and counterpart was thought of in the same terms.

There is evidence of the combination of these two ideas, the king as gardener and the king as shepherd, and more importantly, of their connection to leadership. Jeremiah uses shepherd and farmer to express positions of leadership. In 51:23 he places them in parallel to 'governors' and 'commanders'. In speaking of Babylon as a punishing hammer, Yahweh declares through the prophet:

> With you I break the shepherd (*rōᶜeh*) and his flock,
> With you I break the farmer (*ᵓikkār*) and his team,
> With you I break governors and commanders.[427]

One might also argue that Jeremiah intends the same in 31:23–24: "...The Lord bless you, O habitation of righteousness, O holy hill! Judah and all its cities shall dwell there together, and the farmers (*ᵓikkārîm*) and those who wander with their flocks" (*nāsəᶜû baᶜēder*). The reference to farmers and the shepherds can easily be understood as referring to the leaders of the cities of Judah.[428]

Micah 7:14 appears to combine these two images, again with prominence going to the shepherd image, and with an apparent reference to the primal situation:

> Shepherd your people with your staff,
> the flock of your inheritance,
> who dwell alone in a forest (*yaᶜar*),

---

[426]See, for example, Psalms 29, 47, 93, 95–99.

[427]It is not insignificant that Hebrew *ᵓikkār* is a loanword from Akkadian *ikkaru*. Note also the correspondence between Hebrew *rōᶜeh* and Akkadian *rēᵓû*. It is clear that Jeremiah intends these to signify positions of leadership based on the verses which precede. The groups of parallels are as follows: nations / kingdoms (v. 20); horse and rider/chariot, charioteer (v. 21) ; man and woman/ old and young (22).

[428]In Isa 61:6 the shepherd and the farmer may be significantly juxtaposed with priests and ministers: 'Aliens shall stand and feed your flocks, foreigners shall be your plowmen and vinedressers. But you shall be called the priests of Yahweh. You will be spoken of as the ministers of our God'.

> in the midst of an orchard (***bətôk karmel***),
> let them feed in Bashan and Gilead,
> as in the days of old (***kîmê ʿôlām***).

The picture, a postexilic addition to the prophecy, is of a formerly devastated Israel, now restored to its Eden-like situation. What is essentially the same picture is given by Jeremiah: "I will restore Israel to his pasture, and he shall feed on Carmel and in Bashan, and his desire shall be satisfied on the hills of Ephraim and in Gilead" (50:19). Bashan recalls the primal situation of Eden on two accounts. First, Bashan is noted for its high hills.[429] Second, the reference to *karmel* 'orchard', which appears in both the Micah passage, and the Jeremiah passage recalls the prominent fruit-bearing trees in the Genesis description of the garden. Bashan, in fact, may have even been another cultic place, for we read in Ps 68:16: "O mighty mountain, mountain of Bashan, why do you look with envy at the mount which God desired?"[430] Bashan was well known in antiquity for its fertility, both for its timber and for its grazing plains. In Isa 2:13 and Ezek 27:6 the 'oaks of Bashan' are paralleled by the 'cedars of Lebanon', also associated with the garden of God (Ezek 31:3–9). The cultic orientation of Isa 2:13 is clear in its reference to 'idols', 'high mountains', and 'lofty hills'. Bashan was also associated with the dwelling of the mythical giants, the Rephaim, the race of whom Og, the king, was the only one who remained (Deut 3:11). Thus, there is good cause for Micah's associating Bashan with Eden in 7:14. Yahweh is invoked to be a shepherd in this primal setting, and that primal setting is a *garden*. The two ideas seem to be loosely, yet significantly, associated. Yahweh who is Israel's king, is also Israel's shepherd.[431] Thus, it appears that verses such as Mic 7:14 and Jer 50:19 represent the vestigial remains of the gardener and shepherd royal motifs.

---

[429]The plateau is 2,000 feet above sea level. See J. C. Slayton, "Bashan" *ABD* 1.623–24.

[430]Note also Ps 68:23: 'I will bring them back from Bashan', an apparent reference to Bashan as a rival cultic place.

[431]Bashan, the luxuriant trees, and shepherding is likewise combined in Zech 11:1–3. Note also Deut 32:13, which connects 'herds of Bashan and goats' with 'and the finest of the wheat'.

Ezekiel 28:11–19: The Primal Human Presented as Priest

The imagery presented surrounding the primal human in Ezekiel is priestly. In the oracle preserved in 28:11–19, the primal human on the mountain/garden abode is presented as a priest in the temple. Recognition of the priestly aspect of the primal human in this passage is not a new observation.[432] In chapter 3, we examined the similarity of the language used to describe the primal human (in Ezek 28:11–15) to the topos of the endowing of the king with divine and sacral attributes at his creation. He is distinguished by his possessing the stones of the high priestly breastpiece, an instrument of divination. This imagery is consonant with the description of Adapa, who was described as a priest of the god Ea at Eridu.

Job 15:7–8: The Primal Human Presented in the Language of Prophecy

In Job 15, the language used to describe the primal human is that used with respect to the prophet. The *sôd* of God, which appears in v. 8, is a reference to the ancient Near Eastern divine council. When divine council imagery in the Hebrew Bible involves humanity, that involvement finds expression in the person of the prophet (see Isa 6:8). In Job 15:8, we discovered the primal human 'listening (*šmᶜ*) in the council of (*bəsôd*) God', language which recalls Jeremiah's description of the true prophet, 'standing' in the council of (*bəsôd*) Yahweh to 'perceive', 'hear' (*šmᶜ*), and 'listen to' (*šmᶜ*) his word (Jer 23:18, see also 23:22). It likewise reminds us of the prophet Micaiah ben Imlah's report of a vision, wherein he was present among the divine assembly (1 Kings 22). That the reference in Job 15:8 comes from a work that is generally assigned to the wisdom genre, and that this work reveals no apparent "pro-prophetic agenda," strongly suggests that the image of the primal human in the council of God can be considered one well known to the audience.

---

[432]Note especially recently R. R. Wilson, "The Death of the King of Tyre: The Editorial History of Ezekiel 28," 211–18.

The Primal Human and the Rites of Passage

Incorporation into the sacred world

When we think of the primal human as an intermediary figure, new observations present themselves. A. Van Gennep observed that in rites of passage involving the king (or chief), priests, and magicians, these must be "incorporated into the sacred world" and further that they "cannot function without the operation of the rites of passage."[433] This observation was reiterated and extended in recent decades by V. Turner.[434] The continuity among such intermediary roles as prophet, priest, and king in ancient Israel has long been recognized, and this "sacred incorporation" in van Gennep's theory seems very appropriate to the primal human as we have seen him manifested, as a framework against which some of what we have studied may be understood.

Genesis 2–3

Adam in Genesis is a human incorporated into the sacred world. The account begins with his being extracted from the profane: the dust of the earth. This motif is crucial and, as we have seen, will reappear. He is 'placed' in the garden, the divine abode. In 2:8 we read *wayyāśem šām ʾet-hāʾādām* 'he placed there the human', and in 2:15 *wayyanniḥēhû bəgan ʿēden* 'he placed him in the garden of Eden'. The royal overtones we have observed within the theme of creation and placement within the context of the garden (see above, pp.), suggest that a passage such as Psalm 2 is a reiteration of this notion of a deliberate and, in a sense, official incorporation — reinforcing our interpretation. Yahweh proclaims in 2:6 *nāsaktî malkî ʿal ṣîyôn harqodšî* 'I have set up (or 'installed') my king on Zion, my holy mountain'.[435] Of *hāʾādām*, we were told *wayyanniḥēhû bəgan ʿēden*

---

[433]A. van Gennep, *The Rites of Passage* (Chicago: University of Chicago, 1960), 108–10. Van Gennep postulated three divisions within rites of passage: rites of separation, transition rites, and rites of incorporation (10–11).

[434]*The Ritual Process: Structure and Anti-structure* (Chicago: Aldine, 1969).

[435]We recall that this takes place on the day that he is begotten by God (2:7). To this we may add the reference to Wisdom in Proverbs 8, who was 'brought forth'

lə'obdāh ûləšomrāh '[Yahweh] placed him in the garden of Eden to work and to keep it'.[436] The significance of the incorporation motif in Genesis 2, however, is obscured by the recontextualization of the tradition in a "historical" account.

Ezekiel 28

Ezekiel's primal human is likewise incorporated into the sacred world. In the first oracle (28:1–10), he is a human (*'ādām*) living in the sacred abode, 'the dwelling of the gods in the midst of the sea' (*môšab 'ĕlōhîm... bəlēb yammîm* (v. 2). The text does not reveal the circumstances by which he got there, however. It simply states his presence as a fact. The emphasis on his humanity in conjunction with the statement of his presence (*wə'attâ 'ādām wəlô 'ēl* 'but you are human/ Adam and not God/divine'), however, reveals him not to be original or natural to the setting. He is, in a sense, "out of place," but with a purpose. In the second oracle (28:11–19) Yahweh says, *bəhar qōdeš 'ĕlōhîm hāyîtā* 'you were on the holy mountain of god' (v. 14). As in Genesis, it is a garden referred to by the name Eden; and here we are explicitly told that it is the divine abode: *bəᶜēden gan-'ĕlōhîm* 'Eden, the garden of God' (v. 13). There is a ritual flavor in the statement that the ostensibly sacred articles that are associated with him (high priestly breastpiece, timbrels and pipes) were "established on the day you were created" (28:13). As in Genesis 2, we have a reference to birth (here creation) and being put into a sacred location (see Ps 2:6–7).

Job 15

The allusion to the primal human in Job does not give us explicit details concerning his incorporation into the sacred world. It is clear, however, that the idea is present in the reference that the primal human, "listened" in the *council of God*. Once again, he is "out of place" by virtue of his humanity, but presumably for a sacred reason.

(*ḥôlaltî*, v. 25) set up (*nisssaktî*, v. 23) next to God (*'ehyeh 'eṣlô*, v. 30).

[436]It is difficult to see *wayyanniḥēhû* as a technical term akin to *nāsak*. There may be, however, some significance in its use to describe Yahweh setting Ezekiel in various venues (Ezek 37:1; 40:2)

As we saw in chapter 4, the series of questions in chapter 38 strongly suggest that the primal human was numbered among the "sons of God."[437]

The reason for incorporation is mediation, and it is this notion of incorporation into the sacred world that sets the intermediary apart from the rest of humanity. It is this distinction of place that stands out so clearly in the presentations of the primal human, and provides the basis for the "mythical model" behind the rituals of intermediaries. The intermediary becomes the human who is placed into the divine world, and thereby experiences a direct relationship with God.

Expulsion from the sacred world

According to van Gennep, "the counterpart of initiation rites are the rites of banishment, expulsion, and excommunication — essentially rites of separation and de-sanctification."[438] We should expect no less in ancient Israel, and it is precisely what we find within the primal human mythic complex. Not only do the myths reflect *incorporation* into the sacred world, but they also reflect the converse: expulsion from it or being "profaned" from it, often expressed in terms of death.

Genesis 2–3

In Genesis 3, the primal human is expelled from the sacred world to the profane. Here we can return to van Gennep: "the passage from one social position to another is identified with a territorial passage, such as the entrance into a village or a house, the movement from one room to another, or the crossing of the streets and squares."[439] The primal human in Genesis is driven out from the divine abode, a place where the land produces every kind of fruit-bearing tree, to a place where the earth is obstinate and hostile. Further, this territorial passage is emphasized by mention of the "the cherubim and the flaming sword"

[437]Note that the statement about his birth is placed next to the reference to his listening in the council of God.

[438]*Rites of Passage*, 113.

[439]*Rites of Passage*, 192.

(3:24), established to guard the way back to the tree of life. He returns to dust, devoid of breath and life.

Gen 3:17–19 relates the curse on Adam:

> Cursed is the ground because of you;
> In toil you shall eat of it, for as long as you live.
> Thorns and thistles it will sprout for you,
> and you will eat the plants of the field.
> By the sweat of your brow you will eat bread,
> Until you return to the ground, for from it you were taken.

The curse is executed in 3:23:

> The Lord God sent him from the garden of Eden to work the ground from which he was taken.

There is another observation to be made concerning the curse on the primal human. The expulsion from the garden is an act of profaning, or making or declaring profane. We may also return to the example of Cain, discussed above as a figure reiterating imagery associated with Adam. After Cain's transgression, Yahweh declares his punishment: *wəᶜattā ᵓārûr ᵓāttâ min-hāᵓădāmâ* 'and now you are cursed from the ground' (4:11). It continues in v. 12: ***kî taᶜăbōd ᵓet-hāᵓădāmâ lôᵓ tōsēp tēt-koḥāh lāk*** 'when you work the land, it will not yield its strength to you'. The punishment explicitly recalls that of *hāᵓādām* in Gen 3:17–19, especially v. 18, 'in toil you shall eat of it for as long as you live; thorns and thistles it shall bring forth for you'. In v. 14, the text relates Cain's response to this part of Yahweh's punishment: *hēn gērăštā ᵓōtî hayyôm mēᶜal pənê hāᵓădāmâ ûmippāneykā ᵓessātēr* 'you have driven me today from the face of the ground, and from your face I will be hidden'.

Ezekiel 28

In Ezekiel 28 the primal human is expelled from the sacred world in the second oracle (vv. 11–19). The language explicitly states that he was 'profaned' from the mountain. God says to the primal human in 28:16, *wāᵓeḥallelkā mēhar ᵓĕlōhîm* 'I cast you out as a profane thing (literally 'profaned you') from the mountain of God'. Verses 15–18 reveal the idea behind the expulsion; 28:16 is quite clearly consonant with the imagery concerning Cain's expulsion in Gen 4:14. The text also relates the return to dust, here called "ashes" (see chapter 3). The expulsion of the man in the first oracle (vv. 1–10) is expressed in different terms. The place referred to as *bəlēb yammîm* 'in the heart of the seas', representing the *môšab ʾĕlōhîm* 'dwelling of the gods', the sacred world, becomes in an ironic transformation the place of his death. The transformation is based on the fact that the phrase 'heart of the seas' can refer to the divine dwelling (e.g., the dwelling of El), or to Sheol (see discussion in chapter 5). In this way, that which was sacred (place of the gods) became profaned (place of the slain).

Just as there is a purpose behind the notion of incorporation into the sacred world, so are there reasons for expulsion. In the texts which present an expulsion of the primal human from the world of the sacred, there is always a reason given for this expulsion to the profane. The reasons involve words such as impurity, sin, and violence. The reason, then, is some situation, action, or aspect that is incongruent with the notion of the sacred.

Wisdom plays a particularly central role in this incongruence. In Genesis, the attaining of wisdom by the first humans was presented as problematic. We recall the Adapa myth, where Anu asks, 'why has Ea revealed to an impure man that which pertains to heaven and earth?'.[440] As we discussed in chapter 2, the primal humans' gaining wisdom was also related to the experience at the tower of Babel, where humanity had to be stopped lest they succeed in whatever they consider. In Ezekiel 28, the primal human abused wisdom, which resulted in 'violence'. His actions there are explicitly called 'sin', and related to the profaning of the sacred habitat.

Job 15

The reference to the primal human in Job 15 is brief, and because the speaker assumes audience familiarity with the myth, he does not express the concept of expulsion to the profane. We did see, however,

[440]Fragment B, lines 57–58 (see Heidel, *The Babylonian Genesis*, 151). See above, chapter 2.

that the primal human was introduced within conflict that seemed related to the concept of wisdom.

Eliphaz's comments concerning Job and the primal human (15:14–16) reflect this connection of wisdom with impurity. Within the context of his questions regarding the extent of Job's knowledge, Eliphaz interjects the idea that God places no trust in his holy ones, much less a human (vv. 15,16). Although we are not told of a specific action, such an action is presented as potential and sets God over against humanity. Humankind is impure next to God.

We find in Isa 43:27–28 a similar passage that has been a source of curiosity to scholars. The prophet writes:

27 אָבִיךָ הָרִאשׁוֹן חָטָא
וּמְלִיצֶיךָ פָּשְׁעוּ בִי׃
28 וַאֲחַלֵּל שָׂרֵי קֹדֶשׁ
וְאֶתְּנָה לַחֵרֶם יַעֲקֹב
וְיִשְׂרָאֵל לְגִדּוּפִים׃

27 Your first father sinned,
and your mediators transgressed against me.
28 And *I profaned the princes of the sanctuary*,
and gave Jacob over to destruction,
and Israel to reviling.

Here we have a reference to a 'first father', and to 'mediators'. I would suggest that the prophet is making reference to the phenomenon to which I have been alluding, that the primal human is the paradigm for intermediary offices. The term *mēlîṣ* (v. 27) refers to one who stands between two parties.[441] It appears in Job 33:23 where we read, 'If there is for them (lit. 'him') a messenger (*malʾāk*), a mediator (*mēlîṣ*), one of the thousand to declare (*ləhaggîd*) to humankind what is right for them'. *malʾāk* is, of course, the quintessence of the notion of the intermediary as 'messenger', and *ləhaggîd* (root *ngd*), as we have discussed above (chapter 2), is frequently used to express the revelation of otherwise inaccessible knowledge. The understanding of *mēlîṣ* in both Isa 43:27 and Job 33:23 clearly is of those who stand

[441]In Gen 42:23 it refers to an 'interpreter', one who stands between two linguistic parties, and in 2 Chr 32:31 it refers to an 'envoy' or 'ambassador', one who stands between two political/cultural parties.

between God and humanity. The fact that in Isa 43:27 mediators are parallel to 'your first father' is significant, for as we have seen, the primal human is the paradigmatic one who stands between deity and humanity. The sin of the first father adumbrates the transgression of the mediators. The consequence is "profanation" for the princes of the sanctuary — the same imagery used in Ezek 28:16, and death to Jacob.

A question that remains is, to whom does the designation 'your first father' refer? I believe it is extremely difficult to separate this reference from the notion of the primal human as we have studied it.[442] I should like to go beyond this assertion, however, and conjecture that what we have here is a reference to the patriarch Jacob/Israel *cast as the primal human*. Support for this may be found in the words of Yahweh, spoken through Isaiah in 43:1:

וְעַתָּה כֹּה־אָמַר יְהוָה
בֹּרַאֲךָ יַעֲקֹב וְיֹצֶרְךָ יִשְׂרָאֵל

But now, thus declares Yahweh
He who *created* you, O Jacob
He who *formed* you, O Israel.

The terms used 'create' (*brʾ*) and 'form' (*yṣr*) appear to come straight from the narratives of Genesis 1–2. Just as *hāʾādām* was created (*brʾ*) in Gen 1:27 and formed (*yṣr*) in Gen 2:7, so Jacob/Israel was created and formed in Isa 43:1.

The legitimation of intermediary activity by appeal to such mythic imagery is by no means unparalleled in the ancient world. Recent studies have demonstrated the centrality of the poetic imagination in supporting the imperial cause in ancient Rome.[443] But the texts

---

[442]I assert this despite the fact that some scholars find no evidence for awareness of the Genesis 2–3 account elsewhere in the Hebrew Bible.

[443]Paul Zanker sees modes of creative activity as mirrors of a society that "reflect the state of its values, especially in times of crisis or transition." He asserts "rarely has art been pressed into the service of political power so directly as in the Age of Augustus" (*The Power of Images in the Age of Augustus*, [Ann Arbor: University of Michigan Press, 1990], v). Similarly, Peter White discusses how poets "positioned themselves to address civic or political themes" (*Promised Verse: Poets in the Society of Augustan Rome* [Cambridge, Mass.: Harvard University Press, 1993], ix).

we have studied present both positive and negative ideas about the primal human; thus, we are hardly led to conclude a tradition of uncritical and unconditional support for individuals in such intermediary roles. Instead, the texts reflect a sober and balanced approach to the potentially salubrious or baneful effects of a human serving in role of intermediary. Hence, the regulation of behavior stands out as every bit as important as the function of legitimation. But one need not stray from the ancient Israelite sources themselves for explanation of such a self-conscious expression of checks and balances, for as Herbert Schneidau has observed, one of the outstanding characteristics of Israelite literature, and indeed perhaps its most enduring legacy to western culture, is its relentless reflective self-criticism.[444] The laudatory and condemnatory images associated with the primal human constitute not distinct variant ideas that somehow were later merged. Rather, they reflect the natural and ongoing struggle over the basic and "primal" issues of the definition of "human" and "divine," the nature of ultimate knowledge, its acquisition, and its all too frequent abuse by those who lay claim to it.

---

[444]*Sacred Discontent: The Bible and Western Tradition* (Berkeley: University of California Press, 1977), 14.

Bibliography

Abou-Assaf, Ali, Pierre Bordreuil, and Alan R. Millard. *La statue de Tell Fekherye et son inscription bilingue assyro-araméenne*. Etudes Assyriologiques. Vol. 7. Paris: Editions Recherche sur les civilisations, 1982.

Albright, William F. "Recent Progress in North-Canaanite Research." *BASOR* 70 (1938): 24.

———. *Archaeology and the Religion of Israel*. Baltimore: Johns Hopkins University Press, 1956.

———. "Some Canaanite-Phoenician Sources of Hebrew Wisdom." VTSup 3 (1955): 1–15.

———. "The Early Alphabetic Inscriptions from Sinai and Their Decipherment." *BASOR* 110 (1948): 6–22.

———. *From the Stone Age to Christianity: Monotheism and the Historical Process*. Baltimore: Johns Hopkins University Press, 1957.

Al-Rawi, Farouk N. H., and Jeremy Black. "The Second Tablet of Išum and Erra." *Iraq* 51 (1989): 111–22.

Andreasen, Niels-Erik. "Adam and Adapa: Two Anthropological Characters." *AUSS* 19 (1981): 179–94.

Attridge, Harold, and Robert A. Oden. *Philo of Byblos: The Phoenician History*. The Catholic Biblical Quarterly Monograph Series 9. Washington, D. C.: The Catholic Biblical Quarterly Association of America, 1981.

Auffret, Pierre. "Essai sur la structure littéraire du Psaume 8." *VT* 34 (1984): 257–69.

Bauer, Hans, and Pontus Leander. *Historische Grammatik der hebräischen Sprache des Alten Testaments*. Halle: Niemeyer, 1922. Repr., Hildesheim: Olms, 1962.

Bentzen, Aage. *King and Messiah*. Edited by G. W. Anderson. Oxford: Basil Blackwell, 1970.

———. "King Ideology—'Urmensch'—Troonsbesstijgingsfeest'." *ST* 3 (1950): 148–53.

Bertholet, Alfred. *Hesekiel*. HAT 13. Tübingen: J. C. B. Mohr, 1936.

Bevan, A. A. "The King of Tyre in Ezekiel 28." *JTS* 4 (1903): 500–5.

Bing, J. D. "Adapa and Immortality." *UF* 16 (1984): 53–56.

Bird, Phyllis A. "Male and Female He Created Them: Gen 1:27b in the Context of the Priestly Account of Creation." *HTR* 74 (1981): 129–59.

Blenkinsopp, Joseph. *The Pentateuch: An Introduction to the First Five Books of the Bible*. New York: Doubleday, 1992.

Boer, Pieter A. H. de. "The Counsellor." In *Wisdom in Israel and the Ancient Near East*. Edited by M. Noth and D. W. Thomas. VTSup 3. Leiden: E. J. Brill, 1969. 42–71.

Bonnard, Pierre E. " De la sagesse personnifiée dans l'Ancien Testament à la sagesse en personne dans le Nouveau." In *La Sagesse de l'Ancien Testament*. Edited by M. Gilbert. 117–49. Leuven: Leuven University Press, 1979.

Borger, Rykle. "Die Beschwörungsserie Bīt Mēseri und die Himmelfahrt Henochs." *JNES* 33 (1974): 183–96.

Borsch, Frederick H. *The Son of Man in Myth and History*. New Testament Library. London: SCM Press, 1967.

Bousset, Wilhelm. *Die Religion des Judentums im späthellenistischen Zeitalter*. HNT 21. Tübingen: J. C. B. Mohr, 1926.

———. *Hauptprobleme der Gnosis*. Forschungen zur Religion und Literature des Alten und Neuen Testaments 10. Göttingen: Vandenhoeck und Ruprecht, 1907.

Brandon, Samuel G. F. *Creation Legends of the Ancient Near East*. London: Hodder and Stoughton, 1963.

Brock-Utne, Albert. *Der Gottesgarten: Eine vergleichende religionsgeschichtliche Studie*. Oslo: Dybwad, 1936.

Bruggemann, Walter. "From Dust to Kingship." *ZAW* 84 (1972): 1–18.

Bryce, Glendon E. "Omen-Wisdom in Ancient Israel." *JBL* 94 (1975): 19–37.

Buccellati, Giorgio. "Adapa, Genesis, and the Notion of Faith." *UF* 5 (1973): 61–66.

———. "Wisdom and Not: the Case for Mesopotamia." *JAOS* 101 (1 1981): 35–47.

Budde, Karl. *Die biblische Urgeschichte*. Giessen: J. Ricker, 1883.

———. *Das Buch Hiob übersetzt und erklärt*. Göttingen: Vandenhoeck & Ruprecht, 1896.

Burstein, Stanley. *Berossus the Chaldean*. Sources from the Ancient Near East 1,5. Malibu, Calif.: Undena, 1978.

Cagni, Luigi. *The Poem of Erra*. Sources from the Ancient Near East 1,3. Malibu, Calif.: Undena, 1977.

Cassuto, Umberto. *The Goddess Anath: Canaanite Epics of the Patriarchal Age*. Translated by Israel Abrahams. Jerusalem: Magnes, 1971.

Cazelles, Henri. "La Sagesse de Proverbes 8:22: Peut-elle être considérée comme une hypostase?" In *Trinité et Liturgie*. Edited by A. Triacca and A. Pistoia. 51–57. Roma: C.L.V.-Edizioni Liturgiche, 1984.

Charlesworth, James, ed. *The Old Testament Pseudepigrapha*. 2 vols. New York: Doubleday, 1983.

Cheminant, P. *Les prophéties d'Ezéchiel contre Tyre (XXVI–XXVIII 19)*. Paris: Letouzey et Ane, 1912.

Cheyne, Thomas K. *Traditions and Beliefs of Ancient Israel*. London: Adam and Charles Black, 1907.

Childs, Brevard. "Adam." In *The Interpreter's Dictionary of the Bible*. Edited by G. A. Buttrick. Vol. 1. Nashville: Abingdon, 1962. 42–44.

Christensen, Arthur. *Le premier homme et le premier roi*. Upsala: 1917.

Clay, Albert T. *The Origins of Biblical Traditions: Hebrew Legends in Babylonia and Israel*. Yale Oriental Series 12. New Haven: Yale University Press, 1923.

Clifford, Richard J. *The Cosmic Mountain in Canaan and the Old Testament*. Cambridge, Mass.: Harvard University Press, 1972.

Clines, David J. A. "The Significance of the 'Sons of God' Episode (Genesis 6:1–4) in the Context of the 'Primeval History' (Genesis 1–11)." *JSOT* 13 (1979): 33–46.

Coogan, Michael D. "The Goddess Wisdom—'Where Can She Be Found?': Literary Reflexes of Popular Religion." In *Ki Baruch Hu: Ancient Near Eastern, Biblical, and Judaic Studies in Honor of Baruch A. Levine*. Edited by R. Chazan, W. Hallo, and L. Schiffman. Winona Lake, Ind.: Eisenbrauns, 1999. 203–9.

Cooke, George A. "The Paradise Story of Ezekiel 28." In *Old Testament Essays*. London: Charles Griffin, 1927. 37–45.

———. *A Critical and Exegetical Commentary on the Book of Ezekiel*. ICC. Edinburgh: T. & T. Clark, 1936.

Coppens, Joseph. *La connaissance du bien et du mal et le péché du paradis*. Gembloux: J. Duculot, 1948.

———. "Le messianisme sapiential et les origines littéraires du fils de l'homme daniélique." In *Wisdom in Israel and the Ancient Near East*. Edited by M. Noth and D. W. Thomas. VTSup 3. Leiden: E. J. Brill, 1969. 33–41.

Cornelius, Izak. "Paradise Motifs in the 'Eschatology' of the Minor Prophets andthe Iconography of the Ancient Near East." *JNSL* 14 (1988): 41–83.

Cornill, Carl H. *Das Buch des Propheten Ezechiel*. Leipzig: J. C. Hinrichs, 1886.

Crenshaw, James L. *Old Testament Wisdom*. Atlanta: John Knox, 1981.

———. "The Sage in Proverbs." In *The Sage in Israel and the Ancient Near East*. Edited by J. Gammie and L. J. Perdue. Winona Lake, Ind.: Eisenbrauns, 1990. 205–16.

———. "Wisdom in the OT" In *The Interpreter's Dictionary of the Bible: Supplementary Volume*. Edited by K. Crim. Nashville: Abingdon, 1976. 952–56.

Cross, Frank M. "*ēl*." In *Theological Dictionary of the Old Testament*. Edited by G. J. Botterweck and H. Ringgren. Vol. 1. Grand Rapids: Eerdmans, 1974. 242–61.

———. *Canaanite Myth and Hebrew Epic*. Cambridge, Mass.: Harvard University Press, 1973.

———. "The Council of Yahweh in Second Isaiah." *JNES* 12 (1953): 274–77.

———. "'The Olden Gods' in Ancient Near Eastern Creation Myths," in *Magnalia Dei, The Mighty Acts of God: Essays on the Bible and Archaeology in Memory of G. Ernest Wright*. Edited by F. M. Cross, W. E. Lemke, and P. D. Miller, Jr. Garden City, N.Y.: Doubleday, 1976.

Cross, Frank M., and David Noel Freedman. *Early Hebrew Orthography*. New Haven: American Oriental Society, 1952.

Dalley, Stephanie. *Myths from Mesopotamia*. Oxford: Oxford University Press, 1989.

Day, John. "The Daniel of Ugarit and Ezekiel and the Hero of the Book of Daniel." *VT* 30 (1980): 174–84.

Delitzsch, Franz. *Bibel und Babel*. Leipzig: J. C. Hinrichs, 1902.

Denning-Bolle, Sara. *Wisdom in Akkadian Literature: Expression, Instruction, Dialogue*. Leiden: Ex Oriente Lux, 1992.

Dhorme, Edouard. *A Commentary on the Book of Job*. Translated by Harold Knight. Camden, N.J.: Thomas Nelson and Sons, 1967.

———. *Les religions de Babylonie et d'Assyrie*. Paris: Presses universitaires de France, 1945.

———. *Le Livre de Job*. Études bibliques. Paris: Gabalda, 1926.

Diez Merino, Luis. *Targum de Job: edición principe del Ms. Villa-Amil n. 5 de Alfonso de Zamora*. Biblia Poliglota Complutense. Madrid: Instituto Franciso Suarez, 1984.

Dijk, J. van. "Die Inschriftenfunde." UVB 18 (1962): 39–62.

Dommershausen, W. "*ḥll*." In *Theological Dictionary of the Old Testament*. Edited by G. J. Botterweck and H. Ringgren. Vol. 4. Grand Rapids: Eerdmans, 1974. 409–17.

Dressler, Harold H. P. "The Identification of the Ugaritic Dnil with the Daniel of Ezekiel [reply, B Margalit 30, 361–65]." *VT* 29 (1979): 152–61.

Driver, Godfrey R. "Uncertain Hebrew Words" *JTS* 45 (1944): 13–14.

———. *Canaanite Myths and Legends*. Edinburgh: T. & T. Clark, 1956.

Driver, Samuel R., and G. B. Gray. *A Critical and Exegetical Commentary on the Book of Job*. ICC. Edinburgh: T. & T. Clark, 1921.

Drower, E., and R. Macuch. *A Mandaic Dictionary*. Oxford: Clarendon, 1963.

Dus, Jan. "Melek Sor-Melqart? (Zur Interpretation von Ez 28:11–19)." *ArOr* 26 (1948): 179–85.

Dussaud, René. "Les Phéniciens au Négeb et en Arabie d'après un text de Ras Shamra." *RHR* 108 (1933): 40.

Ebeling, Erich. *Tod and Leben: nach den Vorstellungen der Babylonier*. Berlin, Leipzig: Walter de Gruyter, 1931.

Eichrodt, Walther. *Ezekiel: A Commentary*. OTL. London: SCM Press, 1970.

Eissfeldt, Otto. "Religionsdokument und Religionspoesie, Religionstheorie und Religionshistorie: Ras Schamra und Sanchunjaton, Philo Bybius und Eusebius von Cäsarea." *Theologische Blätter* 17 (1938): 185–97. Repr., *Kleine Schriften* 2 (1963): 130–44.

———. "Zur Frage nach dem Alter der phönizischen Geschichte des Sanchunjaton." *Forschungen und Fortschritte* 14 (1938) 251–52. Repr., *Kleine Schriften* 2 (1963): 127–29.

Eliade, Mircea. *Cosmos and History: The Myth of the Eternal Return*. New York: Harper and Row, 1959.

Emerton, John A. "Origin of Son of Man Imagery." *JTS* 9 (1959): 225–42.

Engnell, Ivan. "Die Urmenschvorstellung und das Alte Testament." *SEA* 22–23 (1957–1958): 265–89.

———. "'Knowledge' and 'Life' in the Creation Story." In *Wisdom in Israel and the*

Ancient Near East. Edited by M. Noth and D. W. Thomas. VTSup 3. Leiden: E. J. Brill, 1969. 103–19.

———. *Studies in Divine Kingship in the Ancient Near East.* Uppsala: Almquist & Wiksell, 1943. Repr., Oxford: Basil Blackwell, 1967.

Erikson, Gösta. "Adam och Adapa." *Svensk T K* 66 (3 1990): 122–28.

Eslinger, Lyle. "A Contextual Identification of the *bene ha*ʾ*elohim* and *benoth ha*ʾ*adam* in Genesis 6:1–4." *JSOT* 13 (1979): 65–73.

Fabry, H. J. "*sôd.*" In *Theologisches Worterbuch zum Alten Testament.* Edited by G. J. Botterweck und H. Ringgren. Vol. 5. Stuttgart: W. Kohlhammer (1973–). 776.

Fensham, F. Charles. "Thunder-Stones in Ugaritic." *JNES* 18 (1959): 273–74.

Feuillet, André. "Le fils de l'homme de Daniel et la tradition biblique." *RB* 60 (1953): 170–203, 321–46.

Fishbane, Michael. "Adam." In *The Encyclopedia of Religion.* Edited by M. Eliade. Vol. 1. New York: Macmillan, 1986. 27–28.

Fohrer, Georg. *Ezechiel.* HAT 13. Tübingen: J. C. B. Mohr (P. Siebeck), 1955.

Foster, Benjamin R. *Before the Muses.* Bethesda, Md.: CDL Press, 1993.

———. "Wisdom and the Gods in Ancient Mesopotamia." *Or NS* 43 (1974): 344–54.

Franke, Judith A. "Presentation Seals of the Ur III/Isin-Larsa Period." *In Seals and Sealings in the Ancient Near East.* Edited by McG. Gibson and R. D. Biggs. Malibu, Calif.: Undena, 1977. 61–66.

Frankfort, Henri. *Kingship and the Gods.* Chicago: University of Chicago Press, 1948.

Frymer-Kensky, Tikva. "The Atrahasis Epic and Its Significance for Our Understanding of Genesis 1–9." *BA* 40 (1977): 147–55.

Gardiner, Alan. *Egypt of the Pharaohs.* Oxford: Oxford University Press, 1961.

Gaster, Theodore. "Ezekiel 28:17." *ET* (1950/51): 124.

———. "Myth and Story." In *Sacred Narrative: Readings in the Theory of Myth.* Edited by A. Dundes. Berkeley and Los Angeles: University of California Press, 1984.

———. *Myth, Legend and Custom in the Old Testament.* New York: Harper and Row, 1969.

———. *Thespis.* New York: Schuman, 1950.

Geertz, Clifford. "Religion as a Cultural System." In *The Interpretation of Cultures.* New York: Basic, 1973.

Gemser, Berend. *Sprüche Salamos.* HAT 16. 2d ed. Tübingen: J. C. B. Mohr, 1963.

Gennep, A. van. *The Rites of Passage.* Chicago: University of Chicago Press, 1960. Originally published as *Rites de passage.* Paris: E. Nourry, 1908.

George, Andrew. *The Epic of Gilgamesh: The Babylonian Epic Poem and Other Texts in Akkadian and Sumerian.* New York: Barnes & Noble, 1999

Geyer, John B. "Mythology and Culture in the Oracles Against the Nations." *VT* 36 (2 1986): 129–45.

Gibson, J. C. L. *Canaanite Myths and Legends*. Edinburgh: T. & T. Clark, 1978.

Ginsberg, Louis. "Adam Kadmon." In *The Jewish Encyclopedia*. Vol. 1. New York and London: Funk and Wagnalls, 1901. 181–83.

Gordis, Robert. *The Book of Job: Commentary, New Translation and Special Studies*. New York: Jewish Theological Seminary of America, 1978.

———. "The Knowledge of Good and Evil in the Old Testament and the Qumran Scrolls." *JBL* 76 (1957): 123–38.

Gordon, C. H. *Ugaritic Literature*. Rome: Pontifical Biblical Institute, 1949.

———. *Ugaritic Manual*. Rome: Pontifical Biblical Institute, 1955.

Görg, Manfred. "Königliche Eulogy: Erwägungen zur Bildsprache in Ps. 8:2." *BN* 37 (1987): 38–47.

Gosse, Bernard. "Ezéchiel 28:11–19 et les détournements de dalédictions." *BN* 44 (1988): 30–38.

———. "Oracles contre les nations et structures comparées des livres d'Isaïe et d'Ezéchiel." *BN* 54 (1990): 19–21.

Gothein, Marie L. *A History of Garden Art*. London and Toronto: J. M. Dent, 1928.

Gowan, Donald. *When Man Becomes God: Humanism and Hybris in the Old Testament*. Pittsburgh Theological Monograph Series 6. Pittsburgh: Pickwick, 1975.

Greenfield, Jonas C. "The Seven Pillars of Wisdom (Prov. 9:1)—A Mistranslation." *JQR* 76 (1985): 13–20.

Gressmann, Hugo. *Der Messias*. Göttingen: Vandenhoeck und Ruprecht, 1929.

———. *Ursprung der israelitsch-jüdischen Eschatologie*. Forschungen zur Religion und Literatur des Alten und Neuen Testaments. Vol. 6. Göttingen: Vandenhoeck und Ruprecht, 1905.

Gunkel, Hermann. *Genesis*. Vol. 1. Göttinger Handkommentar zum alten Testament. Edited by W. Nowack. Göttingen: Vandenhoeck and Ruprecht, 1910.

———. *Schöpfung und Chaos in Urzeit und Endzeit: Eine religionsgeschichliche Untersuchung über Gen 1 und Ap Joh 12*. Göttingen: Vandenhoeck und Ruprecht, 1895.

Haag, Ernst. *Der Mensch am Anfang: Die alttestamentliche Paradiesvorstellung nach Gn 2–3*. Trier: Paulinus-Verlag, 1970.

Haag, H. "*ḥāmās*" In *Theological Dictionary of the Old Testament*. Edited by G. J. Botterweck and H. Ringgren. Vol 4. Grand Rapids: Eerdmans, 1974. 478–86.

Habel, Norman C. "Ezekiel 28 and the Fall of the First Man." *CTM* 38 (1967): 516–24.

———. *The Book of Job: A Commentary*. OTL. London: SCM Press, 1985.

———. "The Symbolism of Wisdom in Proverbs 1–9." *Int* 26 (1972): 131–56.

Hallo, William W. "Antediluvian Cities." *JCS* 23 (1970): 57–67.

———. "'As a Seal upon Thine Arm': Glyptic Metaphors in the Biblical World." In *Ancient Seals and the Bible*. Edited by L. Gorelick and E. Williams-Forte. Malibu, Calif.: Undena, 1983.

———. “On the Antiquity of Sumerian Literature.” *JAOS* 83 (1963): 167–76.

Halpern, Baruch. *The First Historians: The Hebrew Bible and History*. San Francisco: Harper and Row, 1988.

Hamerton-Kelly, Robert. *Pre-existence, Wisdom, and the Son of Man: A Study of the Idea of Pre-Existence in the New Testament*. Society for New Testament Studies Monograph Series 21. Cambridge: Cambridge University Press, 1973.

Hamp, Vinzenz. “ʾēsh.” In *Theological Dictionary of the Old Testament*. Edited by G. J. Botterweck and H. Ringgren. Vol. 1. Grand Rapids: Eerdmans, 1974. 418–28.

Hanson, Paul D. “Rebellion in Heaven, Azazel, and Euhemeristic Heroes in 1 Enoch 6–11.” *JBL* 96 (1977): 195–233.

———. “Zechariah 9 and the Recapitulation of an Ancient Ritual Pattern.” *JBL* 92 (1973): 37–59.

Haran, Menahem. *Temples and Temple Service in Ancient Israel*. Oxford: Oxford University Press, 1978. Repr., Winona Lake, Ind.: Eisenbrauns, 1995.

Harrington, D. J. “Pseudo-Philo: A New Translation and Introduction.” In *The Old Testament Pseudepigrapha*. Edited by James H. Charlesworth. 2 vols. Garden City, N.Y.: Doubleday, 1985. 2.297–377.

Hartman, Louis. “Sin in Paradise.” *CBQ* 20 (1958): 26–40.

Hehn, H. “Zum Terminus ‘Bild Gottes’.” In *Festschrift Eduard Sachau*. Edited by Gotthold Weil (Berlin: G. Reimer, 1915), 36–52.

Heidel, Alexander. *The Babylonian Genesis*. Chicago: University of Chicago Press, 1963.

Hendel, Ronald S. “Of Demigods and the Deluge: Toward an Interpretation of Genesis 6:1–4.” *JBL* 106 (1987): 13–26.

———. “‘The Flame of the Whirling Sword’: A Note on Genesis 3:24.” *JBL* 104 (1985): 671–74.

———. *The Text of Genesis 1–11: Textual Studies and Critical Edition*. New York; Oxford: Oxford University Press, 1998.

Herdner, A. “Remarques sur la ‘déesse ᶜAnat’.” *Revue des études sémitiques-Babyloniaca*, fasc. I (1942–45): 33–49.

Hermann, Johannes. *Ezechiel, übersetzt und erklärt*. KAT. Leipzig: Keichert, 1924.

Hiebert, Theodore. *God of My Victory: The Ancient Hymn in Habakkuk 3*. Harvard Semitic Monographs 38. Atlanta: Scholars Press, 1986.

———. *The Yahwist’s Landscape*. New York and Oxford: Oxford University Press, 1996.

Hillers, Delbert R. “Dust: Some Aspects of Old Testament Imagery.” In *Love and Death in the Ancient Near East: Essays in Honor of Marvin Pope*. Edited by J. H. Marks and R. M. Good. 105–9. Guilford, Conn.: Four Quarters, 1987.

Höffken, Peter F. “Werden und Vergehen der Götter: Ein Beitrag zur Auslegung von Psalm 82.” *TZ* 39 (1983). 129–37.

Hoffner, Harry. *Hittite Myths*. Atlanta: Scholars Press, 1990.

Hölscher, Gustav. *Das Buch Hiob*. HAT 17. Tübingen: J. C. B. Mohr, 1937.

Hooke, Samuel H. *Myth, Ritual, and Kingship: Essays on the Theory and Practice of Kingship in the Ancient Near East and in Israel*. Oxford: Clarendon, 1958.

Hornung, Erik. *Conceptions of God in Ancient Egypt*. Ithaca: Cornell University Press, 1982.

Humbert, Paul. "Démesure et chute dans l'Ancien Testament." In *Hommage à Wilhelm Vischer*, Montpellier: Causse, Graille, Castelnau, 1960.

———. *Études sur le récit du paradis et de la chute dans la Genèse*. Neuchatel: Secrétariat de l'université, 1940.

Hutter, Manfred. "Adam als Gärtner und König (Gen 2:8,15)." *BZ* 30 (1986): 258–62.

Hyatt, James P. "Jeremiah." In *The Interpreter's Bible*. Edited by G. A. Buttrick et al. Vol. 5. Nashville: Abingdon, 1956. 946–1152.

Idel, Moshe. *Kabbala: New Perspectives*. New Haven: Yale University Press, 1988.

Irwin, William A. "Where Shall Wisdom Be Found?" *JBL* 80 (1961): 133–42.

Izre'el, Shlomo. "Some Thoughts on the Amarna Version of Adapa." In *Mésopotamie et Elam. Actes de la XXXVIème Rencontre Assyriologique Internationale*. Mesopotamian History and Environment Occasional Publications 1. Ghent: University of Ghent Press, 1989. 211–20.

Jacobsen, Thorkild. *Harps that Once... Sumerian Poetry in Translation*. New Haven and London: Yale University Press, 1987.

Jahnow, Hedwig. *Das Hebräische Leichenlied*. BZAW 36. Giessen: Alfred Töpelmann, 1923.

Jansen, K. Ludin. *Die Henochgestalt. Eine vergleichende religionsgeschichliche Untersuchung*. Vidensk. akad., Skr., Hist.-filos. kl., 1939.

Jastrow, Morris. "Adam and Eve in Babylonian Literature." *AJSL* 15 (1899): 193–214.

———. *The Religion of Babylonia and Assyria*. Boston: Ginn, 1898.

Jensen, Hans. "The Fall of the King." *SJOT* 1 (1991): 121–47.

Jeppesen, K. "You are a Cherub, but no God! [Ezek 28:1–19]." *SJOT* 1 (1991): 83–94.

Johnson, Aubrey R. *Sacral Kingship in Ancient Israel*. 2d ed. Cardiff: University of Wales Press, 1967.

Johnson, S. E. "Son of Man." In *The Interpreter's Dictionary of the Bible*. Edited by G. A. Buttrick. Vol. 4. Nashville: Abingdon, 1962. 413–20.

Jonge, Marinus de. "Messiah." In *The Anchor Bible Dictionary*. Edited by D. N. Freedman. Vol. 4. New York: Doubleday, 1992. 777–88.

Kapelrud, Arvid S. "Temple Building, A Task for Gods and Kings." *Or NS* 32 (1963): 56–62.

Kayatz, Christa. *Studien zu Proverbien 1–9*. WMANT 22. Neukirchen-Vluyn: Neukirchener Verlag, 1966.

Kikawada, Isaac. "The Double Creation of Mankind in Enki and Ninmah, Atrahasis I 1–351, and Genesis 1–2." *Iraq* 45 (1983): 43–45.

———. "Two Notes on Eve." *JBL* 91 (1972): 33–37.

Kikawada, Isaac, and Arthur Quinn. *Before Abraham Was: The Unity of Genesis 1–11.* Nashville: Abingdon, 1985.

Kilmer, Anne. "Speculations on Umul, the First Baby." In *Kramer Anniversary Volume: Studies in Honor of Samuel Noah Kramer*. Edited by B. L. Eichler et al. AOAT 25. Kevelaer: Butzon & Bercker,1976. 265–70.

———. "The Mesopotamian Counterparts of the Biblical Nĕpīlîm." In *Perspectives on Language and Text: Essays in Honor of Francis I. Andersen's Sixtieth Birthday*. Edited by E. W. Conrad and E. G. Newing. Winona Lake, Ind.: Eisenbrauns, 1987. 39–43.

King, Leonard W. *Babylonian Magic and Sorcery*. London: Luzac, 1896.

Knibb, Michael. *The Ethiopic Book of Enoch*. Oxford: Clarendon, 1978.

Kraeling, Carl H. *Anthropos and Son of Man: A Study in the Religious Syncretism of the Hellenistic Orient*. New York: Columbia University Press, 1927.

Kraeling, Emil G. "Xisouthros, Deucalion and the Flood Traditions." *JAOS* 67 (1947): 177–83.

Kreitzer, Larry. *Prometheus and Adam: Enduring Symbols of the Human Situation*. Lanham, Md.: University Press of America, 1994.

Kutsch, E. "Die Paradieserzählung Gen 2,3 und ihr Verfasser." In *Studien zum Pentateuch*. Edited by G. Braulik. Wien, Freiburg, Basel: Herder, 1977. 9–24.

Kvanvig, Helge S. *Roots of Apocalyptic: The Mesopotamian Background of the Enoch Figure and the Son of Man*. WMANT 61. Edited by G. Bornkamm and G. von Rad. Neukirchen-Vluyn: Neukirchener Verlag, 1988.

Lambdin, Thomas O. "Egyptian Loan Words in the Old Testament." *JAOS* 73 (1953): 145–55.

———. *Introduction to Biblical Hebrew*. New York: Scribner's, 1971.

——— and J. Huehnergard. *Historical Grammar of Biblical Hebrew: Outline*. Unpublished manuscript, 1985.

Lambert, Wilfred G. "A Catalogue of Texts and Authors." *JCS* 16 (1962): 59–77.

———. "A New Look at the Babylonian Background of Genesis." *JTS* 16 (1965): 287–300.

———. "Ancestors, Authors, and Canonicity." *JCS* 14 (1959): 1–14.

———. *Babylonian Wisdom Literature*. Oxford: Clarendon, 1960.

———. "Old Testament Mythology in its Ancient Near Eastern Context." VTSup 40 (1988): 124–43.

Lambert, Wilfred G., and Alan R. Millard. *Atra-ḫasīs: The Babylonian Story of the Flood*. Oxford: Clarendon, 1969.

Langdon, Stephen. "The Legend of Kiskanu." *JRAS* 1928: 843–48.

———. "The Sumero-Babylonian Origin of the Legend of Adam." *ET* 43 (1931–32): 45.

Larsen, Mogens T. *The Old Assyrian City State and Its Colonies*. Copenhagen: Akademisk Forlag, 1976.

Leslau, W. *Ethiopic and South Arabic Contributions to the Hebrew Lexicon*.

Berkeley: University of California Press, 1958.

Levenson, Jon D. *Creation and the Persistence of Evil*. San Francisco: Harper and Row, 1988.

———. *Sinai and Zion*. San Francisco: HarperCollins, 1985.

Lévêque, Jean. "L'argument de la création dans le livre de Job." In *La Création dans L'Orient Ancien*. Edited by L. Derousseaux. 261–99. Paris: Éditions du Cerf, 1987.

Levison, John R. *Portraits of Adam in Early Judaism: From Sirach to 2 Baruch*. Journal for the Study of the Pseudepigrapha Supplement Series 1. Sheffield: Sheffield Academic Press, 1988.

Liagre Böhl, Franz M. T. de. "Das Menschenbild in babylonischer Schau." In *Anthropologie religieuse: L'homme et sa destinée à la lumière de l'histoire des religions*. Edited by C. J. Bleeker. Leiden: E. J. Brill, 1955.

Littmann, E., and M. Höfner. *Wörterbuch der Tigre-Sprache*. Wiesbaden: Franz Steiner, 1962.

Livingstone, David N. "Preadamites: The History of an Idea from Heresy to Orthodoxy." *SJT* 40 (1 1987): 41–66.

Løkkegaard, F. "Some Comments on the Sanchuniathon Tradition." *ST* 8 (1955): 51–76.

Lovejoy, Arthur O., and G. Boas. *Primitivism and Related Ideas in Antiquity: Contributions to the History of Primitivism*. New York: Octagon, 1965.

Mach, Michael. *Entwicklungsstadien des jüdischen Engelglaubens in vorrabbinischer Zeit*. Tübingen: J. C. B. Mohr, 1992.

Machinist, Peter. "Literature as Politics: The Tukulti-Ninurta Epic and the Bible." *CBQ* 38 (1976): 455–82.

Mackay, Cameron. "The King of Tyre." *CQR* 117 (1934): 239–58.

Marböck, Johannes. "Henoch—Adam—der Thronwagen: Zu frühjüdischen pseudepigraphischen Traditionen bei Ben Sira." *BZ* 25 (1981): 103–11.

May, Herbert G. "The King in the Garden of Eden." In *Israel's Prophetic Heritage*. Edited by B. W. Anderson and W. Harrelson. 166–176. New York: Harper, 1962.

Mayer, Werner. R. "Ein Mythos von der Erschaffung des Menschen und des Königs." *Or* 56 (1987): 55–68.

McKane, William. *Proverbs: A New Approach*. London: SCM Press, 1970.

McKenzie, John L. "Mythological Allusions in Ezek. 28:12–18." *JBL* 75 (1956): 322–27.

Meeks, Wayne A. "The Image of the Androgyne: Some Uses of a Symbol in Earliest Christianity." *History of Religions* 13 (1974): 165–208.

Meier, Samuel. "Job 1–2: A Reflection of Genesis 1–3." *VT* 39 (April 1989): 183–93.

———. "Linguistic Clues on the Date and Canaanite Origin of Genesis 2:23–24." *CBQ* 53 (1991): 18–24.

Meinhold, Johannes. *Biblische Urgeschichte*. Bonn: A. Marcus und E. Weber, 1904.

———. *Die Weisheit Israels in Spruch, Sage und Dichtung*. Leipzig: Quelle & Meyer, 1908.

Mertner, Edgar. "Topos und Commonplace." In *Strena Angelica*. Edited by G. Dietrich and F. W. Schultze. Halle: M. Niemeyer, 1956.

Mettinger, Trygve. *King and Messiah: The Civil and Sacral Legitimation of the Israelite Kings*. Coniectanea biblica, Old Testament, 8. Lund: Gleerup, 1976.

Meyer, R. "Zur Sprache von 'Ain Feschcha." *TLZ* 75 [1950]: 721–26.

Michalowski, Piotr. "Adapa and the Ritual Process." *Rocznik Orientalistyczny* 41 (1980): 77–82.

Millard, Alan R., and P. Bordreuil. "A Statue from Syria with Assyrian and Aramaic Inscriptions." *BA* 45 (1982): 135–41.

Millard, Alan R. "The Etymology of Eden." *VT* 34 (1 1984): 103–5.

Miller, Patrick D. "Eridu, Dunnu, and Babel: A Study in Comparative Mythology." *HAR* 9 (1985): 227–51.

———. "Fire in the Mythology of Canaan and Israel." *CBQ* 27 (1965) 256–61.

Moor, Johannes C. de. "East of Eden [Gen 2–3 and Ugaritic Texts KTU 1,100 and 1,107]." *ZAW* 100 (1988): 105–11.

Morgenstern, Julian. "The 'Son of Man' of Daniel 7:13f., A New Interpretation based on 'Two Phase' Solar Religion." *JBL* 80 (1961): 65–77.

———. "The Mythological Background of Psalm 82." *HUCA* 14 (1939): 29–126.

———. "The King-God among the Western Semites and the Meaning of Epiphanes." *VT* 10 (1960): 138–97.

Mowinckel, Sigmund. *He That Cometh*. Translated by G. W. Anderson. New York: Abingdon, 1954.

———. *The Psalms in Israel's Worship*. Translated by D. R. Ap-Thomas. Oxford: Blackwell, 1962.

———. "Urmensch und Königsideologie." *ST* 2 (1948): 71–89.

Muilenburg, James. "The Form and Structure of the Covenantal Formulations." *VT* 9 (1959): 347–65.

———. "The Son of Man in Daniel and the Ethiopic Apocalypse of Enoch." *JBL* 79 (1960): 197–209.

Mullen, E. Theodore. "Divine Assembly." In *The Anchor Bible Dictionary*. Edited by D. N. Freedman. Vol. 2. New York: Doubleday, 1992. 214–17.

———. *The Divine Council in Canaanite and Early Hebrew Literature*. Harvard Semitic Monographs. Chico: Scholars Press, 1980.

Müller, Hans P. "Das Motiv für die Sintflut: die hermeneutische Funktion des mythos und seier Analyse." *ZAW* 3 (1985): 295–316.

———. "Eine neu babylonische Menschenschöpfungserzählung im Licht keilschriftlicher und biblischer Parallelen [VAS 24 no 92]." *Or* 58 (1989): 61–85.

———. "*chāmās*." In *Theological Dictionary of the Old Testament*. Edited by G. J. Botterweck and H. Ringgren. Vol. 4. Grand Rapids: Eerdmans, 1974. 364–84.

Murphy, Roland E. "Wisdom and Creation." *JBL* 104 (1985): 3–11.

———. "Wisdom in the OT." In *The Anchor Bible Dictionary*. Edited by D. N.

Freedman. Vol. 6. New York: Doubleday, 1992. 920–31.

Nickelsburg, George W. E. "Son of Man." In *The Anchor Bible Dictionary*. Edited by D. N. Freedman. Vol. 5. New York: Doubleday, 1992. 137–50.

Niditch, Susan. "The Cosmic Adam: Man as Mediator in Rabbinic Literature." *JJS* 34 (1983): 137–46.

Niditch, Susan, and Robert Doran. "The Success Story of the Wise Courtier: A Formal Approach." *JBL* 96 (1977): 179–93.

Niehr, Herbert. "Götter oder Menschen—eine falsche Alternative: Bemerkungen zu Ps 82." *ZAW* 99 (1987): 94–98.

Oberman, Julian. *Ugaritic Mythology: A Study of Its Leading Motifs*. New Haven: Yale University Press, 1948.

O'Connell, Robert H. "Isaiah 14:4b–23: Ironic Reversal through Concentric Structure and Mythic Allusion [Gilgamesh XI]." *VT* 38 (1988): 406–18.

Oden, Robert A. "Divine Aspirations in Atrahasis and in Genesis 1–11." *ZAW* 93 (1981): 197–216.

———. "Transformations in Near Eastern Myths: Genesis 1–11 and the Old Babylonian Epic of Atrahasis." *Religion* 11 (1981): 21–37.

O'Flaherty, W. D. *The Rig Veda*. London: Penguin, 1981.

Oppenheim, A. Leo. "On Royal Gardens in Mesopotamia." *JNES* 24 (1965): 328–33.

——— et al. *The Assyrian Dictionary of the Oriental Institute of the University of Chicago*. Glückstadt: Augustin, 1956–.

Page, Hugh R. *The Myth of Cosmic Rebellion: A Study of Its Reflexes in Ugaritic and Biblical Literature*. Leiden; New York: E. J. Brill, 1996.

Parpola, Simo. *Letters from Assyrian Scholars to the Kings Esarhaddon and Assurbanipal*. Kevelaer: Butzon & Bercker, 1983.

———. "The Assyrian Tree of Life: Tracing the Origins of Jewish Monotheism and Greek Philosophy." *JNES* 52 (1993): 161–209.

———. *The Standard Babylonian Epic of Gilgamesh*. State Archives of Assyria Cuneiform Texts 1. Helsinki: The Neo-Assyrian Text Corpus Project, University of Helsinki Press, 1997.

Parrot, André. *The Arts of Assyria*. New York: Golden, 1961.

Parys, Michel van. "Création et nature: Le messie et le roi déchu: Une lecture chrétienne du Psaume 8." *Irén* 63 (1990): 5–19.

Pedersen, Johannes. "The Fall of Man." *NTT* 56 (1955): 162–72.

———. "Wisdom and Immortality." In *Wisdom in Israel and the Ancient Near East*. Edited by M. Noth and D. W. Thomas. VTSup 3. Leiden: E. J. Brill, 1955. 238–46.

Perrin, Norman. "Son of Man." *Interpreter's Dictionary of the Bible: Supplementary Volume*. Edited by K. Crim. Nashville: Abingdon, 1976. 833–36.

Pettinato, Giovanni. *Das altorientalische Menschenbild und die sumerischen und akkadischen Schöpfungsmythen*. Heidelberg: Carl Winter Universitätsverlag, 1971.

Picchioni, S. A. *Il poemetto di Adapa*. Budapest: ELTE, 1981.

Pope, Marvin. *El in the Ugaritic Texts*. VTSup 2. Leiden: E. J. Brill, 1955.

———. *Job*. The Anchor Bible. Vol. 15. Garden City, N.Y.: Doubleday, 1973.

———. *Song of Songs*. The Anchor Bible. Vol. 7c. Garden City, N.Y.: Doubleday, 1977.

Prinsloo, Wilhelm. "Isaiah 14:12–15 Humiliation, Hubris, Humiliation." *ZAW* 93 (1981): 432–38.

Rad, Gerhard von. *Das erste Buch Mose. Genesis*. Göttingen: Vandenhoeck und Ruprecht, 1949.

———. "Vom Menschenbild des AT." In *Der alte und der neue Mensch*. Münich: A. Lempp, 1942. 5–23.

Reicke, Bo. "The Knowledge Hidden in the Tree of Paradise." *JSS* 1 (1956): 193–202.

Reiner, Erica. "The Etiological Myth of the Seven Sages." *Or* 30 (1961): 1–11.

Renger, Johannes. "Legal Aspects of Sealing in Ancient Mesopotamia." In *Seals and Sealing in the Ancient Near East*. Edited by McG. Gibson and R. D. Biggs. Malibu, Calif.: Undena, 1977. 75–88.

Richter, Hans Friedemann. "Zur Urgeschichte des Jahwisten." *BN* 34 (1986): 39–57.

Ringgren, Helmer. "*bîn*." In *Theological Dictionary of the Old Testament*. Edited by G. J. Botterweck and H. Ringgren. Vol. 2. Grand Rapids: Eerdmans, 1974. 99–107.

———. "Israel's Place Among the Religions of the Ancient Near East." VTSup 23. Leiden: E. J. Brill, 1972. 1–8.

———. "Remarks on the Method of Comparative Mythology." In *Near Eastern Studies in Honor of W. F. Albright*. Edited by H. Goedicke. Baltimore: Johns Hopkins University Press, 1971. 407–11.

———. "The Impact of the Ancient Near East on Israelite Tradition." In *Tradition and Theology in the Old Testament*. Edited by D. Knight. 31–46. Philadelphia: Fortress, 1977.

———. *Israelite Religion*. Translated by D. E. Green. Philadelphia: Fortress, 1966.

———. *Word and Wisdom: Studies in the Hypostatization of Divine Qualities and Functions in the Ancient Near East*. Lund: Hākan Ohlssons Boktryckeri, 1947.

Rivkin, E. "Messiah, Jewish." In *The Interpreter's Dictionary of the Bible: Supplementary Volume*. Edited by K. Crim. Nashville: Abingdon, 1976. 588–91.

Roberts, J. J. M. *Nahum, Habakkuk, and Zephaniah: A Commentary*. OTL. Louisville: Westminster/John Knox, 1991.

Ross, A. "Proverbs." In The *Expositor's Bible Commentary*. Edited by F. Gaebelein. Vol. 5. Grand Rapids: Zondervan, 1991.

Roux, Georges. "Adapa, le Vent et l'eau." *RA* 55 (1961): 13–33.

Rummel, Stan. "Using Ancient Near Eastern Parallels in Old Testament Study." *BAR* 3 (1977): 8–11.

Rylaarsdam, J. Coert. *Revelation in Jewish Wisdom Literature*. Chicago: The University of Chicago Press, 1946.

Savignac, Jean de. "La sagesse en Proverbes viii.22–31." *VT* 12 (1962): 211–15.

Schencke, Wilhelm. *Die Chokma (Sophia) in der judischen Hypostasenspekulation: Ein Beitrag zur Geschichte der religiosen Ideen im Zeitalter des Hellenismus.* Videnskapsselskapts Skrifter II Hist.-Filos. Klasse 1912. No 6. Kristiania: Utgit for H. A. Benneches Fond, in Kommission bei J. Dybwad, 1913.

Schmidt, Hans. *Die Erzählung von Paradies und Sündenfall.* Tübingen: J. C. B. Mohr, 1931.

Schmidt, Werner. H. *Die Schöpfungsgeschichte der Priesterschrift.* WMANT 17. Neukirchen-Vluyn: Neukirchener Verlag, 1964.

Schneidau, Herbert. *Sacred Discontent: The Bible and Western Tradition.* Berkeley: University of California Press, 1977.

Seitz, Christopher R. "The Divine Council: Temporal Transition and New Prophecy in the Book of Isaiah." *JBL* 109 (1990): 229–47.

Shea, William H. "Adam in Ancient Mesopotamian Traditions." *AUSS* 15 (1977): 27–41.

Smith, James. *The Book of the Prophet Ezekiel.* London: SPCK, 1931.

Smith, Jonathan Z. *To Take Place: Toward Theory in Ritual.* Chicago: University of Chicago Press, 1987.

Smith, Sidney. "An Inscription from the Temple of Sin at Huraidha in the Hadhramawt." *BSOAS* 11 (1945): 451–64.

Speiser, Ephraim A. "Adapa." In *Ancient Near Eastern Texts Relating to the Old Testament.* Edited by J. B. Pritchard. 3d ed. Princeton: Princeton University Press, 1969. 101–3.

———. *Genesis.* The Anchor Bible. Garden City, N.Y.: Doubleday, 1964.

Steenburg, Dave. "The Worship of Adam and Christ as the Image of God." *JSNT* 39 (1990): 95–109.

Steinmann, Jean. *Le prophèt Ézékiel et les débuts de l'exil.* Paris: Editions du Cerf, 1953.

Stronach, David. "The Royal Garden at Parsagadae." *AIO* (1989): 475–502.

Talbert, Charles H. "The Myth of a Descending-Ascending Redeemer in Mediterranean Antiquity." *NTS* 22 (1976): 418–39.

Talon, Philippe. "Le Myth d'Adapa." *SEL* 7 (1990): 43–57.

Terrien, Samuel. *Job.* Commentaire de L'Ancien Testament. Edited by R. Martin-Achard. Vol. 13. Neuchatel: Delachaux et Niestlé, 1963.

Tigay, Jeffrey H. *The Evolution of the Gilgamesh Epic.* Philadelphia: University of Pennsylvania Press, 1982.

Tobin, Thomas H. *The Creation of Man: Philo and the History of Interpretation.* The Catholic Biblical Quarterly Monograph Series. Washington, D. C.: The Catholic Biblical Association of America, 1983.

Trigger, Bruce G., et al. *Ancient Egypt: A Social History.* Cambridge: Cambridge University Press, 1983.

Troje, Luise. "Urmensch." In *Religion in Geschichte und Gegenwart*, 5. Tübingen: J. C. B. Mohr, 1927–1932.

Turner, Victor. *The Ritual Process: Structure and Anti-structure*. Chicago: Aldine, 1969.

Tur-Sinai, N. H. *The Book of Job: A New Commentary*. Jerusalem: Kiryath Sepher, 1957.

Tur-Sinai, N. H. [Torczyner, Harry.] "Presidential Address." *JPOS* 16 (1939): 5.

Van Seters, J. *In Search of History: Historiography in the Ancient World and the Origins of Biblical History*. New Haven: Yale University Press, 1983.

———. "The Creation of Man and the Creation of the King." *ZAW* 101 (1989): 333–42.

Vaux, Roland de. *Genèse*. Paris:L'École biblique de Jérusalem, 1951.

Vawter, Bruce. "Wisdom and Creation." *JBL* 99 (1980): 205–16.

Wallace, Howard N. "Adam." In *The Anchor Bible Dictionary*. Edited by D. N. Freedman. Vol. 1. New York: Doubleday, 1992. 62–64.

———. *The Eden Narrative*. Harvard Semitic Monographs. Atlanta: Scholars Press, 1985.

Waltke, Bruce, and M. O'Connor. *Introduction to Biblical Hebrew Syntax*. Winona Lake, Ind.: Eisenbrauns, 1990.

Waschke, Ernst J. "Untersuchungen zum Menschenbild der Urgeschichte: Ein Beitrag zur alttestamentlichen Theologie." *TLZ* 105 (1980): 795.

Wellhausen, Julius. *Prolegomena to the History of Ancient Israel*. New York: Meridian Books, 1957. Reprint of *Prolegomena to the History of Israel*. Translated by J. Sutherland Black and Allan Enzies, with preface by W. Robertson Smith. Edinburgh: Adam & Charles Black. Translaton of *Prolegomena zur Geschichte Israels*. 2d ed. Berlin: G. Reimer, 1883.

West, Martin L., ed. *Hesiod: Theogony*. Oxford: Clarendon, 1966.

Westermann, Claus. *Genesis 1–11: A Commentary*. Translated by John J. Scullion, S. J. Minneapolis: Augsburg, 1976.

———. "Sinn und Grenze religionsgeschichtlicher Parallelen." In *Forschung am Alten Testament*. Munich: Chr. Kaiser Verlag, 1974. 84–95.

White, Peter. *Promised Verse: Poets in the Society of Augustan Rome*. Cambridge, Mass.: Harvard University Press, 1993.

Whybray, R. N. "Proverbs 8:22–31 and Its Supposed Prototypes." *VT* 15 (1965): 504–14.

———. *The Heavenly Counsellor in Isaiah xl 13–14: A Study of the Sources of the Theology of Deutero-Isaiah*. Society for Old Testament Study Monograph Series 1. Cambridge: Cambridge University Press, 1971.

———. *The Making of the Pentateuch: A Methodological Study*. Sheffield: JSOT Press, 1987.

Widengren, Geo. "Early Hebrew Myths and their Interpretation." In *Myth, Ritual, and Kingship*. Edited by S. Hooke. Oxford: Clarendon, 1958. 149–203.

———. *Psalm 110 och det sakrala kungadömet i Israel.* Uppsala: A.-B. Lundequists, 1941.

———. *The Ascension of the Apostle and the Heavenly Book*: King and Saviour 3. Uppsala: A.-B. Lundequists, 1950.

———. *The King and the Tree of Life in Ancient Near Eastern Religion.* King and Saviour 4. Uppsala: A.-B. Lundequists, 1951.

Wifall, Walter R. "Genesis 6:14: A Royal Davidic Myth?" *BTB* 5 (1975): 294–301.

Wildberger, Hans. "Das Abbild Gottes." *TZ* 21 (1965): 245–59, 481–501.

Williams, Anthony J. "Mythological Background of Ezekiel 28:12–19?" *BTB* 6 (1976): 49–61.

Wilson, Robert R. "The Death of the King of Tyre: The Editorial History of Ezekiel 28." In *Love and Death in the Ancient Near East: Essays in Honor of Marvin H. Pope.* Edited by J. H. Marks and R. M. Good. Guilford, Conn.: Four Quarters, 1987. 211–18.

Wiseman, Donald J. "Mesopotamian Gardens." *AS* 33 (1983): 137–144.

Yaron, Kalman. "The Dirge over the King of Tyre." *ASTI* 3 (1964): 28–57.

Yee, Gail A. "An Analysis of Prov 8:22–31 according to Style and Structure." *ZAW* 94 (1982): 58–66.

Zanker, Paul. *The Power of Images in the Age of Augustus.* Ann Arbor: University of Michigan Press, 1988.

Ziegler, Joseph. *Ezechiel.* Göttingen: Vandenhoeck und Ruprecht, 1952.

Zimmerli, Walther. *Das Menschenbild des alten Testament.* Theologische Existenz Heute. Münich: C. Kaiser, 1949.

———. *Ezekiel.* 2 vols. Translated by James D. Martin. Hermeneia. Philadelphia: Fortress, 1983.

Zimmern, Heinrich. *Die Keilinschriften und das Alte Testament.* Berlin: Reuther & Reichard, 1902.

Zurro Rodríguez, E. "El hápax *ʾabrēk* (Gn 41,43)." *Estudios Bíblicos* 49 (1991): 265–69.

Index of Textual Citations

1. Hebrew Bible

Genesis

2. New Testament

3. Apocrypha

4. Pseudepigrapha

5. Qumran Writings

6. Rabbinic Texts

7. Ugaritic Texts

8. Aramaic Texts

9. Akkadian and Sumerian Texts

10. Greek Texts

11. Other

Printed in the United States
By Bookmasters